21世纪高等学校物联网专业系列教材

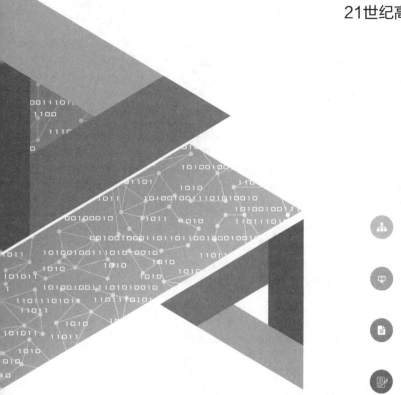

物联网专业英语教程

第2版

张强华 司爱侠 编著

清华大学出版社
北京

内 容 简 介

本书的目的在于切实提高物联网行业人士的专业英语能力。

本书以物联网专业应用实际为依据，采集难度适中、覆盖面广的实用性和前瞻性材料组成单元。课文包括物联网简介、物联网体系结构、M2M、传输介质、物联网协议、物联网传感器、网络体系结构、网络拓扑、组网硬件、网络交换、无线传感器网络、无线网络、RFID、物联网开发平台、物联网安全、用于物联网的区块链技术、物联网数据分析及行业物联网。阅读材料包括物联网实际应用、物联网网络、物联网设备、因特网、路由器、物联网与智慧城市、NFC、物联网常用工具、物联网的云计算技术及人工智能物联网。每个单元包括 Text A 及 Text B，课文包括了基础知识和基本概念；New Words，给出课文中出现的新词，读者由此可以积累基本的专业词汇；Phrases，给出课文中的常用词组；Abbreviations，给出课文中出现的、业内人士必须掌握的缩略语；Exercises，针对课文练习及扩展练习，巩固学习效果、扩展能力；Reading Material，可进一步扩宽读者的视野。

本书可作为高等院校的专业英语教材，也可作为培训班教材，还可供从业人员自学。

版权所有，侵权必究。举报：010-62782989，beiqinquan@tup.tsinghua.edu.cn。

图书在版编目（CIP）数据

物联网专业英语教程 / 张强华，司爱侠编著. 2版. -- 北京：清华大学出版社，2024.12. -- (21世纪高等学校物联网专业系列教材). -- ISBN 978-7-302-67766-6

Ⅰ. TP393.4；TP18

中国国家版本馆CIP数据核字第20241AB879号

责任编辑：安　妮
封面设计：刘　键
责任校对：王勤勤
责任印制：刘　菲

出版发行：清华大学出版社
网　　址：https://www.tup.com.cn, https://www.wqxuetang.com
地　　址：北京清华大学学研大厦A座
邮　编：100084
社 总 机：010-83470000
邮　购：010-62786544
投稿与读者服务：010-62776969，c-service@tup.tsinghua.edu.cn
质 量 反 馈：010-62772015，zhiliang@tup.tsinghua.edu.cn
课 件 下 载：https://www.tup.com.cn, 010-83470236

印 装 者：三河市春园印刷有限公司
经　　销：全国新华书店
开　　本：185mm×260mm　　印　张：14.75　　字　数：353千字
版　　次：2015年9月第1版　　2024年12月第2版　　印　次：2024年12月第1次印刷
印　　数：1～1500
定　　价：49.00元

产品编号：105799-01

前言
FOREWORD

物联网是继计算机、互联网之后信息产业发展的第三次浪潮。它通过智能感知、识别技术与普适计算，结合网络化应用，创造性地拓展和变革了众多行业。其影响的深度与广度都十分巨大，物联网人才需求旺盛。因此，我国数百所高校开设了相关专业。由于物联网各组成部分均处于高速发展之中，国际化特征尤为明显，从业人员必须提高专业英语水平，以便及时获得最新、最先进的专业知识。从某种意义上说，专业英语的水平决定了专业技能的水平。因此，几乎所有开设物联网专业的高校都开设了相应的专业英语课程。

本书以物联网专业应用实际为依据，采集难度适中、覆盖面广的实用性和前瞻性材料，课文包括物联网简介、物联网体系结构、M2M、传输介质、物联网协议、物联网传感器、网络体系结构、网络拓扑、组网硬件、网络交换、无线传感器网络、无线网络、RFID、物联网开发平台、物联网安全、用于物联网的区块链技术、物联网数据分析及行业物联网。阅读材料包括物联网实际应用、物联网网络、物联网设备、因特网、路由器、物联网与智慧城市、NFC、物联网常用工具、物联网的云计算技术及人工智能物联网。每个单元包括Text A 及 Text B，课文包括了基础知识和基本概念；New Words，给出课文中出现的新词，读者由此可以积累基本的专业词汇；Phrases，给出课文中的常用词组；Abbreviations，给出课文中出现的、业内人士必须掌握的缩略语；Exercises，针对课文练习及扩展练习，巩固学习效果、扩展能力；Reading Material，可进一步扩宽读者的视野。本书提供了词汇总表的电子版，供读者记忆单词和长期查询之用，可扫描下方二维码下载。

扫码下载词汇总表

我们向教师提供教学支持资料，包括教学课件、教学大纲、电子教案、习题答案、模拟试卷和课程资料，可从清华大学出版社官网下载。

因编者水平有限，书中不当之处在所难免，望大家不吝赐教。让我们共同努力，使本书成为一部"符合学生实际、切合行业实况、知识实用丰富、严谨开放创新"的优秀教材。

编者
2024 年 8 月

目录
CONTENTS

Unit 1 1

  Text A   Introduction to IoT 1
      New Words 4
      Phrases 6
      Abbreviations 7
      Analysis of Difficult Sentences 7
      Exercises 8
  Text B   Introduction to IoT Architecture 10
      New Words 13
      Phrases 14
      Exercises 14
      Reading Material 16
  参考译文 18

Unit 2 21

  Text A   What Is M2M? 21
      New Words 23
      Phrases 25
      Abbreviations 25
      Analysis of Difficult Sentences 26
      Exercises 26
  Text B   Types of Transmission Media 28
      New Words 31
      Phrases 32
      Abbreviations 33
      Exercises 33
      Reading Material 35

参考译文 37

Unit 3 40

Text A　IoT Protocols 40
　　New Words 44
　　Phrases 45
　　Abbreviations 46
　　Analysis of Difficult Sentences 46
　　Exercises 47

Text B　IoT Sensors 50
　　New Words 53
　　Phrases 55
　　Abbreviations 56
　　Exercises 56
　　Reading Material 58
　　参考译文 61

Unit 4 65

Text A　Network Architecture 65
　　New Words 68
　　Phrases 70
　　Abbreviations 71
　　Analysis of Difficult Sentences 71
　　Exercises 72

Text B　Network Topology 75
　　New Words 80
　　Phrases 81
　　Abbreviations 82
　　Exercises 82
　　Reading Material 84
　　参考译文 86

Unit 5 89

Text A　Networking Hardware 89
　　New Words 93
　　Phrases 94
　　Abbreviations 95
　　Analysis of Difficult Sentences 95
　　Exercises 96

Text B　Network Switch 98
　　　　New Words 104
　　　　Phrases 105
　　　　Abbreviations 107
　　　　Exercises 107
　　　　Reading Material 108
　　参考译文 112

Unit 6 116

　　Text A　Wireless Sensor Network 116
　　　　New Words 120
　　　　Phrases 121
　　　　Abbreviations 122
　　　　Analysis of Difficult Sentences 122
　　　　Exercises 123
　　Text B　WiFi 126
　　　　New Words 130
　　　　Phrases 131
　　　　Abbreviations 131
　　　　Exercises 132
　　　　Reading Material 133
　　参考译文 137

Unit 7 140

　　Text A　How Wireless Networks Work 140
　　　　New Words 143
　　　　Phrases 144
　　　　Abbreviations 145
　　　　Analysis of Difficult Sentences 145
　　　　Exercises 146
　　Text B　RFID 148
　　　　New Words 151
　　　　Phrases 152
　　　　Abbreviations 153
　　　　Exercises 153
　　　　Reading Material 154
　　参考译文 158

Unit 8 160

　　Text A　Top 5 IoT Development Platforms 160

New Words 164
　　Phrases 165
　　Abbreviations 166
　　Analysis of Difficult Sentences 166
　　Exercises 167
　Text B　Programming Languages for IoT 169
　　New Words 172
　　Phrases 173
　　Abbreviations 173
　　Exercises 173
　　Reading Material 175
　参考译文 179

Unit 9 183

　Text A　IoT Security 183
　　New Words 186
　　Phrases 188
　　Abbreviations 189
　　Analysis of Difficult Sentences 189
　　Exercises 190
　Text B　Blockchain for IoT 193
　　New Words 196
　　Phrases 197
　　Exercises 198
　　Reading Material 199
　参考译文 202

Unit 10 205

　Text A　IoT Data Analytics 205
　　New Words 208
　　Phrases 209
　　Analysis of Difficult Sentences 210
　　Exercises 211
　Text B　IIoT 214
　　New Words 218
　　Phrases 218
　　Abbreviations 219
　　Exercises 219
　　Reading Material 220
　参考译文 226

Unit 1

Text A

扫码听音频

Introduction to IoT

1. What Is IoT (Internet of Things)?

IoT is a network of physical objects or people called "things" that are embedded with software, electronics, network, and sensors for the purpose of collecting and exchanging data with other devices and systems over the internet. These devices range from ordinary household objects to sophisticated industrial tools.

IoT makes virtually everything "smart" by improving every aspect of our life with the power of data collection, AI algorithm, and networks. The thing in IoT can also be a person with a diabetes monitor implant, an animal with tracking devices, etc.

2. How Does IoT Work?

The entire IoT process starts from the devices themselves, such as smartphones, smartwatches, televisions, washing machines, and other electronic devices, which help you to communicate with the IoT platform.

The following are four fundamental components of an IoT system:
- Sensors/devices: sensors or devices are a key component that helps you to collect live data from the surrounding environment. All this data may have various levels of complexities. The data could be from a simple temperature monitoring sensor, or it may be in the form of the video feed.

A device may have various types of sensors and it performs multiple tasks apart from sensing. For example, a mobile phone is a device which has multiple sensors like GPS, and cameras.
- Connectivity: all the collected data is sent to a cloud infrastructure. The sensors should be connected to the cloud using various mediums of communications. These

communication mediums include mobile or satellite networks, Bluetooth, WiFi, WAN, etc.
- Data processing: once that data is collected and it gets to the cloud, the software performs processing on the gathered data. This process can be just checking the temperature, reading on devices like heaters. However, it can sometimes also be very complex like identifying objects, using computer vision[1] on video.
- User interface[2]: the information needs to be available to the end user in some way, which can be achieved by triggering alarms on their phones or sending them notification through email or text message. The user sometimes might need an interface which actively checks their IoT system. For example, the user has a camera installed in his home. He wants to access video recording and all the feeds with the help of a Web server[3].

However, it's not always one-way communication. Depending on the IoT application and complexity of the system, the user may also be able to perform an action which may create cascading effects. For example, if a user detects any changes in the temperature of the refrigerator, with the help of IoT technology the user should able to adjust the temperature with the help of their mobile phone.

There are also cases where some actions can be performed automatically. By establishing and implementing some predefined rules, the entire IoT system can adjust the settings automatically and no human has to be physically present.

Also in case any intruders are sensed, the system can generate an alert not only to the owner of the house but to the concerned authorities.

3. IoT Applications

IoT solutions are widely used in numerous companies across industries. Some most common IoT applications are given below(see Table 1.1).

Table 1.1 Common IoT applications

Application type	Description
Smart thermostats	Smart thermostats can help you to save resource on heating bills by knowing your usage patterns.
Connected cars	IoT helps automobile companies handle billing, parking, insurance, and other related stuff automatically.

---

1 Computer vision is a field of artificial intelligence (AI) enabling computers to derive information from images, videos and other inputs.

2 The user interface is the point at which human users interact with a computer, website or application. The goal of effective UI is to make the user's experience easy and intuitive, requiring minimum effort on the user's part to receive the maximum desired outcome.

3 On the hardware side, a Web server is a computer that stores Web server software and a website's component files (for example, HTML documents, images, CSS stylesheets, and JavaScript files). A Web server connects to the Internet and supports physical data interchange with other devices connected to the Web.

Application type	Description
Activity trackers	Activity trackers can help you capture heart rate pattern, calorie expenditure, activity levels, and skin temperature on your wrist.
Smart outlets	With smart outlets, you can remotely turn any device on or off. They also allow you to track a device's energy level and get custom notifications directly into your smartphone.
Parking sensors	Parking sensors can help users to identify the real-time availability of parking spaces on their phone.
Connected health care system	A connected health care system facilitates real-time health monitoring and patient care. It helps in improved medical decision-making based on patient data.
Smart city	Smart city offers all types of use cases which include traffic management, water distribution, waste management, etc.
Smart home	Smart home integrates the devices in your home together, which include smoke detectors, home appliances, light bulbs, windows, door locks, etc.
Smart supply chain[1]	Smart supply chain helps you in real time tracking of goods while they are on the road, or getting suppliers to exchange inventory information.

4. Challenges of IoT

At present IoT is faced with many challenges, such as:
- Insufficient testing and updating.
- Concern regarding data security and privacy.
- Software complexity.
- Data volumes and interpretation.
- Integration with AI and automation.
- Devices require a constant power supply, which is difficult.
- Interaction and short-range communication.

5. Advantages of IoT

Key benefits of IoT technology are as follows:
- Technical optimization: IoT technology helps a lot in improving technologies and making them better. For example, with IoT, a manufacturer is able to collect data from various car sensors. the manufacturer analyzes them to improve its design and make them more efficient.
- Improved data collection: traditional data collection[2] has its limitations and its design is for passive use. IoT facilitates immediate action on data.
- Reduced waste: IoT offers real-time information leading to effective decision making & management of resources. For example, if a manufacturer finds an issue in multiple car

---

1　A supply chain is a network of companies and people that are involved in the production and delivery of a product or service. The components of a supply chain include producers, vendors, warehouses, transportation companies, distribution centers, and retailers.

2　Data collection is the systematic process of gathering observations or measurements in research. It can be qualitative or quantitative.

engines, he can track the manufacturing plan of those engines and solves this issue within the manufacturing belt.
- Improved customer engagement: IoT allows you to improve customer experience by detecting problems and improving the process.

6. Disadvantages of IoT

Now, let's see some of the disadvantages of IoT:
- Security: IoT technology creates an ecosystem of connected devices. However, during this process, the system may offer little authentication control despite sufficient security measures.
- Privacy: the use of IoT exposes a substantial amount of personal data in extreme detail, without the user's active participation. This creates lots of privacy issues.
- Flexibility: there is a huge concern about the flexibility of an IoT system. It is mainly about integrating with another system as there are many diverse systems involved in the process.
- Complexity: the design of the IoT system is also quite complicated. Moreover, its deployment and maintenance is also not very easy.
- Compliance: IoT has its own set of rules and regulations. However, because of its complexity, the task of compliance is quite challenging.

7. Cautions When Applying IoT

The following cautions should be taken when applying IoT.
- Design products with reliability and safety in mind.
- Use strong authentication and security protocols.
- Disable non-essential services.
- Ensure internet is well managed, and IoT management hubs & services are secured.
- Design more energy-efficient algorithms so that the system remains active longer.

New Words

network	['netwɜːk]	n. 网络
embed	[ɪm'bed]	v. 把……嵌入
internet	['ɪntənet]	n. 互联网
smart	[smɑːt]	adj. 智能的
standard	['stændəd]	n. 标准
		adj. 标准的，合格的
device	[dɪ'vaɪs]	n. 设备，装置
computer	[kəm'pjuːtə]	n.（电子）计算机
tablet	['tæblət]	n. 平板电脑
data	['deɪtə]	n. 数据

algorithm	['ælgərɪðəm]	n.	算法
monitor	['mɒnɪtə]	n.	屏幕，显示器
		vt.	监控
process	['prəʊses]	n.	过程，流程
		vt.	加工；处理
fundamental	[,fʌndə'mentl]	adj.	基础的，基本的
complexity	[kəm'pleksəti]	n.	复杂性
feed	[fiːd]	vi.	馈送，进入
		vt.	向……提供
infrastructure	['ɪnfrəstrʌktʃə]	n.	基础设施
medium	['miːdɪəm]	n.	介质，媒介物
communication	[kə,mjuːnɪ'keɪʃn]	n.	通信
satellite	['sætəlaɪt]	n.	卫星；人造卫星
		v.	通过通信卫星传播
bluetooth	['bluːtuːθ]	n.	蓝牙（一种无线传输技术）
software	['sɒftweə]	n.	软件
gather	['gæðə]	vt.	收集；聚集
email	['iːmeɪl]	n.	电子邮件
		vt.	给……发电子邮件
interface	['ɪntəfeɪs]	n.	界面；接口
video	['vɪdɪəʊ]	n.	录像
		adj.	视频的
perform	[pə'fɔːm]	v.	执行
action	['ækʃn]	n.	行动，活动
detect	[dɪ'tekt]	vt.	检测，发现
adjust	[ə'dʒʌst]	v.	调整
automatically	[,ɔːtə'mætɪkli]	adv.	自动地
intruder	[ɪn'truːdə]	n.	闯入者，侵入者；干扰者
sense	[sens]	vt.	感到；检测出
		n.	感觉；识别力
generate	['dʒenəreɪt]	vt.	形成，造成，产生
numerous	['njuːmərəs]	adj.	很多的，许多的
application	[,æplɪ'keɪʃn]	n.	应用，适用
thermostat	['θɜːməstæt]	n.	恒温器
pattern	['pætn]	n.	模式
handle	['hændl]	vi.	处理，操作
capture	['kæptʃə]	vt. & n.	捕获
remotely	[rɪ'məʊtli]	adv.	远程地

real-time	['riːlˈtaɪm]	adj. 实时的
availability	[əˌveɪləˈbɪləti]	n. 有效；有益；可利用性
concept	[ˈkɒnsept]	n. 观念，概念
decision-making	[dɪˈsɪʒnˈmeɪkɪŋ]	n. 决策
inventory	[ˈɪnvəntri]	n. 库存，存货清单
insufficient	[ˌɪnsəˈfɪʃnt]	adj. 不足的，不够的
update	[ˌʌpˈdeɪt]	vt. 更新
	[ˈʌpdeɪt]	n. 更新
interpretation	[ɪnˌtɜːprɪˈteɪʃn]	n. 解释，说明；翻译
optimization	[ˈɒptəmaɪˈzeɪʃən]	n. 最佳化，最优化
analyze	[ˈænəlaɪz]	vt. 分析；分解
engagement	[ɪnˈɡeɪdʒmənt]	n. 参与度
ecosystem	[ˈiːkəʊsɪstəm]	n. 生态系统
authentication	[ɔːˌθentɪˈkeɪʃn]	n. 身份验证；认证；证明
privacy	[ˈpraɪvəsɪ]	n. 隐私
substantial	[səbˈstænʃl]	adj. 大量的
deployment	[dɪˈplɔɪmənt]	n. 部署；调度
maintenance	[ˈmeɪntənəns]	n. 维护；维修
compliance	[kəmˈplaɪəns]	n. 合规性
rule	[ruːl]	n. 规则；规章
regulation	[ˌreɡjuˈleɪʃn]	n. 规章；规则
reliability	[rɪˌlaɪəˈbɪləti]	n. 可靠性
protocol	[ˈprəʊtəkɒl]	n. 协议
disable	[dɪsˈeɪbl]	vt. 使无效

Phrases

dumb device	哑设备
tracking device	跟踪器
electronic device	电子设备，电子器件
surrounding environment	周围环境
temperature monitoring sensor	温度监测传感器
multiple task	多任务
cloud infrastructure	云基础设施
communication medium	通信介质
data process	数据处理
computer vision	计算机视觉
user interface	用户界面
end user	终端用户

text message	短信，文本消息
Web server	网络服务器
be able to	能，会
cascading effect	级联效应
predefined rule	预定义规则
calorie expenditure	热量消耗
base on	基于
smart city	智慧城市
smart home	智能家居
smoke detector	烟雾探测器
data volume	数据量
integration with ...	与……集成
data collection	数据收集，数据汇集
customer experience	客户体验
personal data	个人数据
management hub	管理中心

Abbreviations

IoT (Internet of Things)	物联网
AI (Artificial Intelligence)	人工智能
GPS (Global Position System)	全球定位系统
WiFi (Wireless Fidelity)	无线保真

Analysis of Difficult Sentences

[1] IoT is a network of physical objects or people called "things" that are embedded with software, electronics, network, and sensors for the purpose of collecting and exchanging data with other devices and systems over the internet.

本句中，called "things"是一个过去分词短语，作定语，修饰和限定 physical objects or people。that are embedded with software, electronics, network, and sensors for the purpose of collecting and exchanging data with other devices and systems over the internet 是一个定语从句，修饰和限定 physical objects or people。for the purpose of collecting and exchanging data with other devices and systems over the internet 在定语从句中作目的状语。

[2] The entire IoT process starts from the devices themselves, such as smartphones, smartwatches, televisions, washing machines, and other electronic devices, which help you to communicate with the IoT platform.

本句中，such as smartphones, smartwatches, televisions, washing machines, and other electronic devices 是对 the devices 的举例说明。which help you to communicate with the IoT platform 是一个定语从句，修饰和限定 the devices 进行补充说明。

[3] Also in case any intruders are sensed, the system can generate an alert not only to the

owner of the house but to the concerned authorities.

本句中，in case any intruders are sensed 是一个条件状语从句，修饰主句的谓语 can generate。

英语中，in case 可以引导条件状语从句，也可以引导目的状语从句。例如：

In case the program has any problems, let me know as soon as possible.

如果程序有什么问题，请尽快告诉我（条件状语从句）。

Our manager asked me to write down the product model in case I forget.

我们经理让我把产品型号写下来以免忘了（目的状语从句）。

[4] Smart city offers all types of use cases which include traffic management, water distribution, waste management, etc.

本句中，which include traffic management, water distribution, waste management, etc.是一个定语从句，修饰和限定 use cases。

[5] IoT offers real-time information leading to effective decision making & management of resources.

本句中，leading to effective decision making & management of resources 是一个现在分词短语，作定语，修饰和限定 real-time information。该短语可以扩展为一个定语从句：which leads to effective decision making & management of resources。

Exercises

【EX.1】Answer the following questions according to the text.

1. What does IoT stand for? What is it?
2. How does IoT make virtually everything "smart"?
3. What are the four fundamental components of an IoT system?
4. What do the communication mediums used to connect sensors to the cloud include?
5. How can the entire IOT system adjust the settings automatically and no human has to be physically present?
6. What does IoT help automobile companies to do ?
7. What does smart city offer?
8. What is the first challenge mentioned that IoT is faced with at present?
9. How does IoT allow you to improve customer experience?
10. Is there a huge concern about the flexibility of an IoT system? What is it mainly about and why?

【EX.2】Translate the following terms or phrases from English into Chinese and vice versa.

1.	algorithm	1.	_____
2.	compliance	2.	_____
3.	deployment	3.	_____
4.	device	4.	_____

5.	engagement	5.	
6.	inventory	6.	
7.	customer experience	7.	
8.	data collection	8.	
9.	management hub	9.	
10.	smart home	10.	
11.	用户界面	11.	
12.	*adj.* 智能的	12.	
13.	*vt.* 感到；检测出	13.	
14.	*adv.* 远程地	14.	
15.	*n.* 协议	15.	

【EX.3】 Translate the following sentences into Chinese.

1. A firewall provides an essential security blanket for your computer network.
2. More and more people are using the internet.
3. This computer is popular for its good design and ease of use.
4. This device helps make virtual reality a more usable and accessible technology.
5. Performance analysis and optimization are essential and necessary.
6. The company has spent thousands of pounds updating their computer systems.
7. Network security technology mainly includes authentication, encryption, access control, auditing and so on.
8. For the practical and applied system, the reliability is very important.
9. The operator has to be able to carry out routine maintenance of the machine.
10. Their first task will be to set up a communications system.

【EX.4】 Complete the following passage with appropriate words in the box.

development	embedded	technologies	tagged	check
applications	tracked	computing	concept	network

IoT is a scenario in which every thing has a unique identifier and the ability to communicate over the Internet or a similar wide area network (WAN).

　　The __1__ for an IoT are already in place. Things, in this context, can be people, animals, servers, __2__, shampoo bottles, cars, steering wheels, coffee machines, park benches or just about any other random item that comes to mind. Once something has a unique identifier, it can be __3__, assigned a uniform resource identifier (URI) and monitored over a __4__ network. IoT is an evolutionary outcome of the trend towards ubiquitous __5__, a scenario in which processors are __6__ in everyday objects.

Although the ___7___ wasn't named until 1999, IoT has been in ___8___ for decades. The first Internet appliance was a Coke machine at Carnegie Melon University in the early 1980s. Programmers working several floors above the vending machine wrote a server program that ___9___ how long it had been since a storage column in the machine had been empty. The programmers could connect to the machine over the Internet, ___10___ the status of the machine and determine whether or not there would be a cold drink awaiting them, should they decide to make the trip down to the machine.

【EX.5】 Translate the following passage into Chinese.

What does IoT mean?

IoT is a computing concept that describes a future where everyday physical objects will be connected to the Internet and will be able to identify themselves to other devices. The term is closely identified with RFID as the method of communication, although it could also include other sensor technologies, other wireless technologies, QR codes, etc.

IoT is significant because an object that can represent itself digitally becomes something greater than the object existed by itself. No longer does the object relate just to you, but now it is connected to objects around it, data from a database, etc. When many objects act in unison, they are referred to as having "ambient intelligence".

Most of us think about being connected in terms of computers, tablets and smartphones. IoT describes a world where just about anything can be connected and communicate in an intelligent fashion. In other words, with the Internet of Things, the physical world is becoming one big information system.

Text B

Introduction to IoT Architecture

1. Definition of IoT Architecture

In the case of IoT, architecture refers to the planned network involving devices, cloud technology and network structure, which can enable communication among IoT devices.

2. Layers of IoT Architecture

The four types of models which explain the architecture of Internet of Things solutions include 3-layer architecture, 4-layer architecture, 5-layer architecture and 7-layer architecture.

2.1　3-layer Architecture

The 3-layer architecture for the Internet of Things applications has been the most common model for defining IoT application design. Here is an outline of the elements in the three-layer Internet of Things architecture.

2.1.1　Perception Layer

The perception layer in the basic architecture of IoT includes the sensors used for IoT solutions and networks. It serves as the core component for IoT functionality by offering access to data, which can be collected from different sensors on IoT-connected devices. The perception layer also includes actuators, which respond according to changes in their environment.

2.1.2 Network Layer

The network layer is another crucial component in the domain of IoT as it explains the movement of data in the IoT network. Network layer is crucial to the IoT architecture for establishing connectivity among the different devices alongside sending data to the relevant backend services.

2.1.3 Application Layer

Application layer serves as the user-oriented layer of an IoT solution, such as the smartphone app for controlling IoT devices in smart homes.

2.2 5-layer Architecture

2.2.1 Perception Layer

Perception layer represents the physical devices in the IoT networks that "perceive" data required for processing. IoT sensors in different applications, such as health monitoring systems, autonomous vehicles, security systems, and lighting systems, provide an example of a perception layer in IoT.

2.2.2 Transport Layer

Transport layer is responsible for transferring the data collected to the edge or cloud centers for processing. It utilizes internet gateways for transferring data from the perception layer, i.e., sensors, to the processing layer.

2.2.3 Processing Layer

The capability of IoT solutions to deliver value depends on performance in processing data collected from IoT devices. New IoT architectures utilize artificial intelligence and machine learning to create value through the analysis of data. For example, IoT devices can not only record fluctuations in temperature but also use AI to create alerts by comparing temperatures to predefined benchmarks.

2.2.4 Application Layer

The application layer deals with the user interface of IoT solutions. Most of the processing tasks in IoT architectures do not require human intervention. Application layer helps administrators in managing IoT devices and creating rules for operations of the IoT network. In addition, the application layer also involves the design of service-level agreements[1] for IoT systems and networks.

2.2.5 Business Layer

The 5-layer IoT architecture also includes a promising addition in the form of a business

---

1 A service-level agreement (SLA) defines the level of service you expect from a vendor, laying out the metrics by which service is measured, as well as remedies or penalties should agreed-on service levels not be achieved. It is a critical component of any technology vendor contract.

layer. It focuses on the transformation of IoT data into business intelligence[1], which can drive effective decision-making approaches. Most of the work of the business layer depends on reports alongside live dashboards for facilitating business intelligence.

2.3 4-layer Architecture

2.3.1 Devices

Devices can refer to sensors or actuators which generate data or respond to the environment. The data generated by devices is translated into a digital format for transmission to the internet gateway. With the exception of time-sensitive use cases, the data from IoT devices is transferred in a raw state to gateways.

2.3.2 Internet Gateways

The internet gateways are responsible for receiving data from different stages, followed by preprocessing before sending it to the cloud. Generally, the internet gateway can be a standalone or integrated device that can interact with sensors and transfer information to the internet.

2.3.3 Edge Computing

You can rely on transferring data to the edge of the cloud to facilitate faster routes of data processing. As a result, you can ensure faster data analysis and identify the areas which need urgent attention. The edge computing layer focuses primarily on time-sensitive operations and also involves preprocessing. The advantages of preprocessing help in limiting the amount of data sent to the cloud.

2.3.4 Cloud or Data Center

In the last stage of the IoT solution, you will have to store data for processing at a later stage. The cloud or data center encompasses the application and business layers in the basic architecture of IoT solutions.

2.4 7-layer Architecture

2.4.1 Perception Layer

The first step of core IoT infrastructure functioning comprises a number of special devices equipped with sensors that take on the role of intermediators between analog and digital worlds. These devices have various forms and sizes, and are separated into three distinct categories.

2.4.2 Connectivity Layer

The next stage is the connection between all the devices, sensors and services that function together. These can be connected in two different ways:
- Direct connection (TCP, UDP/IP).
- Gateway connection (using modules that can translate, encrypt and decrypt data).

2.4.3 Fog Computing[2] Layer

Fog (edge) computing layer is aimed at quick data storage near its sources. This layer

---

1 Business intelligence (BI) is a set of strategies and technologies for analyzing business information and transforming it into actionable insights that inform strategic and tactical business decisions.

2 Fog computing is a decentralized infrastructure that places storage and processing components at the edge of the cloud, where data sources such as application users and sensors exist.

makes it possible to analyze and transform large amounts of data in real time near its original sources. It is a time- and resource-saving approach that results in quicker system response and overall improved functioning.

2.4.4 Processing Layer

While the previous layers are used for constant data modification, this is not the case at the processing layer. This layer is responsible for accumulating, storing and processing information received from the fog computing layer.

2.4.5 Application Layer

At this point, data undergoes analysis by software programs and apps with the aim to provide the answers needed to make well-thought-out business decisions. Currently a great number of apps exist that differ in their functions, implemented operation systems[1] and technologies.

2.4.6 Business Layer

The business layer helps companies adjust their business strategies and make more thoughtful plans based on the right data. It enables them to dive deeper into the issues that need to be resolved as well as their causes, and predict possible consequences.

2.4.7 Security Layer

Big companies/suppliers of IoT devices install high-level security by default, but the ability to add additional protection features once the system is in use is of paramount importance. There are major levels of device security: device-level, connection-level and cloud-level.

3. Conclusion

The explanations for different types of IoT architecture models show the guidelines for designing IoT solutions. Interestingly, each model presents a distinct explanation for the architecture of Internet of Things networks.

New Words

outline	['aʊtlaɪn]	n. 大纲，提纲，要点
element	['elɪmənt]	n. 元素；要素
actuator	['æktʃʊeɪtə]	n. 执行器，执行机构
backend	['bækend]	n. 后端
user-oriented	['juːzə'ɔːrɪəntɪd]	adj. 用户导向的，面向用户的
record	['rekɔːd]	n. 记录
fluctuation	[ˌflʌktʃʊ'eɪʃn]	n. 波动，涨落，起伏
predefine	['priːdɪ'faɪn]	vt. 预定义；预先确定
benchmark	['bentʃmɑːk]	n. 基准，参照
intervention	[ˌɪntə'venʃn]	n. 介入，干涉，干预

---

1 An operating system (OS) is system software that manages computer hardware and software resources, and provides common services for computer programs.

transformation	[ˌtrænsfə'meɪʃn]	n. 转换；变化
dashboard	['dæʃbɔːd]	n. 仪表板，仪表盘
preprocess	[priː'prəʊses]	vt. 预处理，预加工
standalone	['stændəˌləʊn]	adj. 单独的，独立的
urgent	['ɜːdʒənt]	adj. 紧急的，急迫的
encompass	[ɪn'kʌmpəs]	vt. 围绕，包围
equip	[ɪ'kwɪp]	vt. 装备，配备
intermediator	[ˌɪntə'miːdɪeɪtə]	n. 中介，中间人
distinct	[dɪ'stɪŋkt]	adj. 明显的，清楚的
gateway	['geɪtweɪ]	n. 网关
accumulate	[ə'kjuːmjəleɪt]	v. 积累，堆积
paramount	['pærəmaʊnt]	adj. 最高的，至上的；最重要的
guideline	['gaɪdlaɪn]	n. 指导方针；指导原则

Phrases

network structure	网络结构
perception layer	感知层
core component	核心组件
application layer	应用层
physical device	物理设备
health monitoring system	健康监护系统
cloud center	云中心
processing layer	处理层
service-level agreement	服务水平协议，服务等级协议
business layer	业务层
business intelligence	商业智能
edge computing layer	边缘计算层
be separated into	被分成
connectivity layer	连接层
fog computing	雾计算
original source	初始源
operation system	操作系统
security layer	安全层

Exercises

【EX.6】Answer the following questions according to the text.

1. What does architecture refer to in the case of IoT?
2. How many types of models which explain the architecture of IoT solutions are there? What are they?

3. What does the perception layer in the basic architecture of IoT include?
4. What does application layer serve as?
5. What are the elements in the 5-layer IoT architecture?
6. What is transport layer responsible for?
7. What are the internet gateways responsible for?
8. What does the edge computing layer focus primarily on?
9. What does the business layer help companies do?
10. What are the major levels of device security mentioned in the passage?

【EX.7】 Translate the following terms or phrases from English into Chinese and vice versa.

1.	actuator	1.	
2.	dashboard	2.	
3.	equip	3.	
4.	gateway	4.	
5.	intervention	5.	
6.	preprocess	6.	
7.	transformation	7.	
8.	application layer	8.	
9.	business intelligence	9.	
10.	fog computing	10.	
11.	感知层	11.	
12.	安全层	12.	
13.	服务水平协议，服务等级协议	13.	
14.	n. 记录	14.	
15.	n. 元素；要素	15.	

【EX.8】 Translate the following sentences into Chinese.
1. The report outlined possible uses for the new device.
2. In order to make it safe, the element is electrically insulated.
3. It is also possible to authenticate each user at the backend server.
4. Telecom operators have gradually changed their development point from network capacity to user-oriented services.
5. Accurate records must be kept at all times.
6. The new fast computer is now the benchmark in the computer industry.
7. You have to preprocess the data to make the data mining more effective.
8. The laboratory has been newly equipped.
9. No internet gateways are needed for the program's operation.

10. This guideline might also be appropriate for other types of products.

Reading Material

Applications of IoT

The ubiquity[1] of IoT is a fact of life thanks to its adoption by a wide range of industries. Let us learn about the applications of IoT in various industries.

1. IoT Applications in Agriculture

For indoor planting, IoT makes monitoring and management of micro-climate conditions a reality, which in turn increases production. For outside planting, devices using IoT technology can sense soil moisture[2] and nutrients[3], in conjunction with weather data, better control smart irrigation[4] and fertilizer[5] systems.

2. IoT Applications in Consumer Use

For the private citizen, IoT devices in the form of wearables and smart homes make life easier.

Smart homes take care of things like activating environmental controls so that your house is at peak comfort[6] when you come home. Security is made more accessible as well, with the consumer having the ability to control appliances and lights remotely, as well as activating a smart lock to allow the appropriate people to enter the house even if they don't have a key.

3. IoT Applications in Healthcare

First and foremost, wearable IoT devices let hospitals monitor their patients' health at home, thereby reducing hospital stays while still providing up to the minute real-time information that can save lives. In hospitals, smart beds keep the staff informed of the available beds, thus shortening the waiting time for vacant beds. Putting IoT sensors on critical equipment means fewer breakdowns and increased reliability.

4. IoT Applications in Manufacturing

RFID and GPS technology can help a manufacturer track a product from its start on the factory floor to its placement in the destination store, the whole supply chain from start to finish. These sensors can gather information on travel time, product condition, and environmental conditions that the product is subjected to.

Sensors mounted on those same machines can also track the performance of the machine, predicting when the unit will require maintenance, thereby preventing costly breakdowns.

1　ubiquity [juːˈbɪkwəti] *n*. 到处存在，（同时的）普遍存在
2　soil moisture: 土壤湿度
3　nutrient [ˈnjuːtriənt] *n*. 养分，养料
4　irrigation [ˌɪrɪˈgeɪʃn] *n*. 灌溉；水利
5　fertilizer [ˈfɜːtəlaɪzə] *n*. 肥料，化肥
6　comfort [ˈkʌmfət] *n*. 舒适

5. IoT Applications in Retail

IoT technology can bring many benefits to the retail industry. Online[1] and in-store shopping sales figures and information gleaned from IoT sensors can be used to warehouse automation. Much of this relies on RFIDs, which are already in heavy use worldwide.

Speaking of customer engagement[2], IoT helps retailers target customers based on past purchases. Equipped with the information provided through IoT, a retailer can craft a personalized promotion[3] for their loyal customers, thereby eliminating the need for costly mass-marketing promotions that don't stand as much of a chance of success. Much of these promotions can be conducted through the customers' smartphones, especially if they have an app for the appropriate store.

6. IoT Applications in Transportation

Most people have heard about the progress being made with self-driving cars. But that's just one bit of the vast potential in the field of transportation. The GPS[4], which is another example of of IoT, is being utilized to help transportation companies plot faster and more efficient routes for trucks hauling freight, thereby speeding up delivery times.

City planners can also use that data to help determine traffic patterns, parking space demand, and road construction and maintenance.

7. IoT Applications in Wearables

The fitness bracelet[5] can monitor calorie consumption, walking distance, heartbeats per minute, blood oxygen levels, etc.These IoT mostly come in the form of wristbands[6]/watches. However, they can also appear as earbuds, clip-on[7] devices, or smart fabric.

Other wearables include virtual glasses and GPS tracking belts. These small and energy-efficient devices equipped with sensors and software collect and organize data about users.

8. IoT Applications in Traffic Monitoring

A major contributor to the concept of smart cities, the Internet of Things is beneficial in vehicular traffic management in large cities. Using mobile phones as sensors to collect and share data from our vehicles via applications like Google Maps is an example of using IoT. It informs about the traffic conditions of the different routes, estimated arrival time and the distance from the destination while contributing to traffic monitoring.

Traffic pattern analysis gives commuters a perfect idea of what might happen during peak hours. Thus, they will be better prepared to avoid traffic by being aware of possible alternatives[8].

1　online [ˌɒn'laɪn] *adj.* 在线的，联网的，联机的
2　engagement [ɪn'geɪdʒmənt] *n.* 参与
3　promotion [prə'məʊʃn] *n.* 促销
4　Global Position System: 全球定位系统
5　fitness bracelet: 健身手环
6　wristband ['rɪstbænd] *n.* 腕套，腕带
7　clip-on ['klɪpˌɔn] *adj.* 可用夹子夹住的
8　alternative [ɔːl'tɜːnətɪv] *adj.* 替代的，备选的

9. IoT Applications in Smart Grid and Energy Saving

A smart grid is a holistic solution employing information technology to reduce electricity waste and cost, improving electricity efficiency, economics, and reliability.

The establishment[1] of bidirectional[2] communication between the end user and the service provider allows substantial value to fault detection, decision making, and repair thereof. It also helps users monitor their consumption patterns and adopt the best ways to reduce energy expenditure.

10. IoT Applications in Smart Cities

IoTs have already made several cities more efficient such that they need fewer resources and have increased energy efficiency.

Sensors in different capacities throughout the city are combined to complete various tasks such as managing the traffic, handling waste management, optimizing streetlights[3], saving water, monitoring energy expenditure[4], creating smart buildings, and more.

参考译文

Text A 参考译文　物联网简介

1. 什么是物联网（IoT）？

物联网是一个由被称为"物"的物理对象或人组成的网络，这些物理对象或人嵌入了软件、电子设备、网络和传感器，用于通过互联网与其他设备和系统收集和交换数据。这些设备的范围是从普通的家用物品到复杂的工业工具。

物联网利用数据收集、人工智能算法和网络的力量改善我们生活的各方面，从而使几乎一切变得"智能"。物联网中的"物"还可以是植入糖尿病监测仪的人、带有跟踪设备的动物等。

2. 物联网如何运行？

整个物联网流程从设备本身开始，如智能手机、智能手表、电视、洗衣机和其他帮助你与物联网平台进行通信的电子设备。

以下是物联网系统的4个基本组成部分。

- 传感器/设备：传感器或设备是帮助你从周围环境收集实时数据的关键部件。这些数据可能具有不同程度的复杂性。数据可以来自简单的温度监控传感器，也能以视频输入的形式提供。

1　establishment [ɪˈstæblɪʃmənt] *n.* 建立
2　bidirectional [ˌbaɪdəˈrekʃənl] *adj.* 双向的
3　streetlight [ˈstriːtlaɪt] *n.* 街灯
4　expenditure [ɪkˈspendɪtʃə] *n.* 耗费；花费，支出

设备可能具有各种类型的传感器，并且除了感测之外还执行多种任务。例如，移动电话是一种具有多个传感器（如 GPS 和摄像头）的设备。
- 连接性：所有收集到的数据都被发送到云基础设施。应使用各种通信介质将传感器连接到云端。这些通信介质包括移动或卫星网络、蓝牙、WiFi、WAN 等。
- 数据处理：一旦数据被收集并发送到云端，软件就会对收集到的数据进行处理。这个过程可以只是检查温度，或读取加热器的温度。然而，它有时也可能非常复杂，如使用计算机视觉技术来识别视频中的物体。
- 用户界面：信息需要以某种方式提供给最终用户，实现方法可以是通过触发手机上的警报，也可以通过电子邮件或短信向用户发送信息。用户有时可能需要一个主动检查其物联网系统的界面。例如，用户在家里安装了摄像头，他希望借助网络服务器能够访问录制的视频和推送的所有信息。

然而，这并不总是单向的通信。根据物联网应用和系统的复杂性，用户还可以执行可能产生级联效应的操作。例如，如果用户检测到冰箱温度有任何变化，借助物联网技术，用户就能够通过手机调节温度。

在一些情况下，物联网会自动执行某些操作。通过建立和实施一些预定义的规则，整个物联网系统可以自动调整设置，而无须人在场。

此外，如果检测到任何入侵者，系统不仅可以向房主发出警报，还可以向有关当局发出警报。

3. 物联网应用

物联网解决方案被广泛应用于各行各业的众多公司。表 1.1 给出了一些最常见的物联网应用。

表 1.1　物联网常用应用

应用类型	描　　述
智能恒温器	智能恒温器能够了解你的使用模式，帮助你节省取暖费用
联网汽车	物联网帮助汽车公司自动处理计费、停车、保险和其他相关事务
活动追踪器	活动追踪器能够帮助你捕获手腕上的心率模式、热量消耗、活动水平和皮肤温度
智能插座	有了智能插座，你可以远程打开或关闭任何设备。它们还允许你跟踪设备的电量等级并将自定义通知直接发送到你的智能手机
停车传感器	停车传感器能够帮助用户在手机上实时识别可用的停车位
连接健康	连接的医疗保健系统有利于实时健康监测和患者护理。它有助于根据患者数据改进医疗决策
智慧城市	智慧城市提供各种类型的用例，包括交通、供水、废物的管理等
智能家居	智能家居把你家中的设备集成在一起，包括烟雾探测器、家用电器、灯泡、窗户、门锁等
智慧供应链	智慧供应链帮助你实时跟踪货物运输状况，或让供应商交换库存信息

4. 物联网的挑战

目前物联网面临着诸多挑战，例如：
- 测试和更新不足。

- 对数据安全和隐私的担忧。
- 软件复杂性。
- 数据量和解释。
- 与人工智能和自动化的集成。
- 设备需要持续供电,这是很困难的。
- 交互和短距离通信。

5. 物联网的优势
物联网技术的主要优势如下:
- 技术优化:物联网技术对于改进技术并使其变得更好十分有益。例如,通过物联网,制造商能够从各种汽车传感器收集数据,对其进行分析以改进设计并提高效率。
- 改进数据收集:传统的数据收集有其局限性,其设计是被动使用的。物联网有助于对数据采取即时操作。
- 减少浪费:物联网提供实时信息,从而实现有效的决策和资源管理。例如,如果制造商发现多个汽车发动机存在问题,那么他可以跟踪这些发动机的制造计划并在制造厂内解决该问题。
- 提高客户参与度:物联网使你能够通过检测问题和改进流程来改善客户体验。

6. 物联网的缺点
现在,让我们看看物联网的一些缺点:
- 安全性:物联网技术创建了一个互联设备的生态系统。然而,在此过程中,尽管有足够的安全措施,但是系统可能提供的身份验证控制很少。
- 隐私:物联网的使用会在没有用户积极参与的情况下暴露大量极其详细的个人数据。这会产生很多隐私问题。
- 灵活性:人们对物联网系统的灵活性非常关注。它主要涉及与另一个系统的集成,因为该过程涉及许多不同的系统。
- 复杂性:物联网系统的设计也相当复杂。它的部署和维护也不是很容易。
- 合规性:物联网有自己的一套规则和法规。然而,由于其复杂性,合规任务颇具挑战性。

7. 物联网应用的注意事项
应用物联网时应注意以下事项:
- 设计产品时考虑可靠性和安全性。
- 使用强大的身份验证和安全协议。
- 禁用非必要服务。
- 确保互联网得到良好管理,并且物联网管理中心和服务是安全的。
- 设计更节能的算法,使系统活动的时间更长。

Unit 2

Text A

What Is M2M?

M2M refers to data communications between two or more machines. M2M is most commonly translated as machine to machine but sometimes it is translated as man to machine, machine to man and others. Cellular telephone service providers use public wireless networks to accomplish M2M, telemetry or telematics[1].

Most often, M2M systems are task-specific, meaning that a given system is purpose-built for just one specific device, or a very restricted class of devices in an industry. This is one of the indicators of the M2M market which is still in its infancy, as a unified intercommunication standard has yet to evolve. Functions are duplicated—each purpose-built system repeats many functions already implemented in similar systems.

Wireless M2M is machine to machine communication through wireless technologies such as CDMA.

1. Where Can M2M Be Used?

M2M provides benefits to many individuals, companies, communities and organizations in the public and private sectors across various industries. The following are a few "emerging segments" that are gaining growth benefits through leveraging M2M solutions:

- Telecommunication—IP internetworking / wireless WAN / mobile learning.
- Manufacturing—supply chain / inventory management / factory automation.
- Light industry—sensor monitoring / remote access control / utilities.
- Transportation—asset tracking / logistics management.

---

1 Telematics typically is any integrated use of telecommunications and informatics, also known as ICT (Information and Communications Technology, 信息与通信技术).

- Retail—point-of-sale / kiosk / digital content signage.
- Telematics—aftermarket / in-vehicle solutions.
- E-business / m-business solutions.
- Real estate—building automation.
- Security—video surveillance.

2. What Are the Benefits of M2M?

These are just some of the benefits that are gained through leveraging M2M to solve business challenges:

- Cost-effective preventive maintenance and quality of service.
- Fast response through outsourcing troubleshooting.
- Centralized service support and data management.
- On-going revenues throughout product lifecycle.
- Increased revenues from minimized downtime.
- Remote diagnostics.

3. What Makes Up an M2M Solution?

Usually an M2M solution is made up of several components:

- Device(s) or sensor(s) to collect data and/or monitor changes.
- An application and database to process the data sent and received.
- A central server to send and / or receive data transmitted by the device(s) or sensor(s).
- Connectivity (either fixed line or wireless) to connect the device or sensor to a central server.
- A modem to allow data exchange between the device(s) or sensor(s) and the central server.
- An application to ensure security of the data transmitted and to monitor and manage the connectivity to the device or sensor network.

4. What Are the Benefits of Wireless M2M?

4.1 Flexibility

Devices that are wirelessly connected to a network are not limited to a physical location. This gives you the flexibility to move them should you need to, for example if a vending machine is not getting much footfall, you can simply place it elsewhere. Devices in remote locations where it's difficult to run cables can also be attached to a wireless network quickly and easily.

4.2 Mobility

Mobile devices can connect into a network, which is not possible with a fixed line network, allowing data communication with all devices.

4.3 Access to Information

Wireless networks deliver realtime information to mobile devices enabling information to be received or sent whenever and wherever it is required, so organizations have access to live information for effective and fast decisionmaking.

4.4 Independent Network

A wireless network can be quickly and easily deployed into a building or location without the need for integration with the existing fixed line network, thereby delivering an independent secure network.

4.5 Speed

A wireless network can be deployed much more quickly than a fixed network as no cabling is needed between devices. This allows devices to be active more quickly and can provide cost savings.

4.6 Cost

There are cost savings related to the speed of deployment, the elimination of cabling costs, and the reduction in communication costs, as CDMA cellular is more cost effective than PSTN/fixed line.

4.7 How Do Wireless M2M Machines Communicate?

M2M wireless solutions use wireless modems to communicate using always-on CDMA cellular allowing it to communicate data immediately, and at much higher speeds. CDMA airtime/data usage is billed according to the amount of data used and not the amount of time the connection lasts.

New Words

accomplish	[ə'kʌmplɪʃ]	vt. 完成，达到，实现
telemetry	[tə'lemətri]	n. 遥感勘测，自动测量记录传导
telematics	[ˌtelɪ'mætɪks]	n. 信息通信业务，远程信息处理
purpose-built	['pɜːpəs 'bɪlt]	adj. 为特定目的建造的
specific	[spə'sɪfɪk]	n. 细节
		adj. 详细而精确的，明确的，特殊的
restrict	[rɪ'strɪkt]	vt. 限制，约束，限定
industry	['ɪndəstri]	n. 工业；行业
indicator	['ɪndɪkeɪtə]	n. 指标；指示器
infancy	['ɪnfənsɪ]	n. 幼年
intercommunication	[ˌɪntəkəˌmjuːnɪ'keɪʃn]	n. 双向（或多向）通信
evolve	[ɪ'vɒlv]	v. (使)发展，(使)进展，(使)进化
function	['fʌŋkʃn]	n. 功能，作用
duplicate	['djuːplɪkeɪt]	vt. 使成双，使加倍；复制
	['djuːplɪkət]	n. 复制品；复印件
		adj. 复制的，副本的；成对的，二倍的
implement	[ɪmplɪmənt]	v. 实施，实现

emerging	[ɪˈmɜːdʒɪŋ]	adj.	新兴的，不断出现的，涌现的
gain	[ɡeɪn]	vt.	获得，得到
leverage	[ˈliːvərɪdʒ]	vt.	起杠杆作用
		n.	杠杆
retail	[ˈriːteɪl]	n.	零售
		adj.	零售的
	[rɪˈteɪl]	v.	零售
automation	[ˌɔːtəˈmeɪʃn]	n.	自动控制，自动操作
utility	[juːˈtɪləti]	n.	效用，有用
logistic	[lɒˈdʒɪstɪk]	adj.	物流的，后勤的
kiosk	[ˈkiːɒsk]	n.	亭子
aftermarket	[ˈæftəˌmɑːkɪt]	n.	零件市场
surveillance	[sɜːˈveɪləns]	n.	监视，监督
preventive	[prɪˈventɪv]	adj.	预防性的
centralize	[ˈsentrəlaɪz]	vt.	集聚，集中
revenue	[ˈrevənjuː]	n.	收入，税收
downtime	[ˈdaʊntaɪm]	n.	停工期
component	[kəmˈpəʊnənt]	n.	成分
		adj.	组成的，构成的
database	[ˈdeɪtəbeɪs]	n.	数据库，资料库
ensure	[ɪnˈʃʊə]	v.	确保，保证
connectivity	[ˌkɒnekˈtɪvəti]	n.	连通性
server	[ˈsɜːvə]	n.	服务器
modem	[ˈməʊdem]	n.	调制解调器
exchange	[ɪksˈtʃeɪndʒ]	vt.	交换
flexibility	[ˌfleksəˈbɪləti]	n.	弹性，适应性，灵活性
footfall	[ˈfʊtfɔːl]	n.	客流量
attach	[əˈtætʃ]	vt.	缚上，系上，贴上
mobility	[məʊˈbɪləti]	n.	活动性，移动性，机动性
live	[laɪv]	adj.	活的，生动的，精力充沛的
effective	[ɪˈfektɪv]	adj.	有效的，被实施的
independent	[ˌɪndɪˈpendənt]	adj.	独立的，不受约束的
deploy	[dɪˈplɔɪ]	v.	部署，展开，配置
integration	[ˌɪntɪˈɡreɪʃn]	n.	综合，集成
deliver	[dɪˈlɪvə]	vt.	递送，交付
elimination	[ɪˌlɪmɪˈneɪʃn]	n.	排除，除去，消除
reduction	[rɪˈdʌkʃn]	n.	减少，缩影，变形，缩减量
immediately	[ɪˈmiːdɪətli]	adv.	立即，马上，直接地

Phrases

refer to	指；涉及；查阅；有关
cellular telephone	移动电话
public wireless network	公共无线网络
in one's infancy	初期，早期
mobile learning	移动学习
supply chain	供应链
inventory management	库存管理
factory automation	工厂自动化
light industry	轻工业
sensor monitoring	传感器检测
remote access control	远程访问控制
asset tracking	资产跟踪
logistics management	物流管理
digital content signage	数字标牌
in-vehicle solutions	车载解决方案
m-business solutions	移动商务解决方案
real estate	房地产
building automation	楼宇自动化
video surveillance	视频监控
preventive maintenance	预防性维修，定期检修
outsourcing troubleshooting	外包的故障排除
data management	数据管理
product lifecycle	产品生命周期
remote diagnostics	远程诊断
vending machine	自动贩卖机
realtime	实时
mobile device	移动设备
decision making	决策，判定
cost saving	节约成本
communication cost	通信成本

Abbreviations

M2M (Machine to Machine)	机器对机器
CDMA (Code Division Multiple Access)	码分多址
WAN (Wide Area Network)	广域网
PSTN (Public Switched Telephone Network)	公用电话交换网

Analysis of Difficult Sentences

[1] This is one of the indicators of the M2M market which is still in its infancy, as a unified intercommunication standard has yet to evolve.

本句中，which is still in its infancy, as a unified intercommunication standard has yet to evolve 是一个定语从句，修饰和限定 the M2M market。在该从句中，as a unified intercommunication standard has yet to evolve 是一个原因状语从句，修饰谓语 is still in its infancy。in its infancy 的意思是"处于萌芽阶段，在初始阶段"。

[2] These are just some of the benefits that are gained through leveraging M2M to solve business challenges.

本句中，that are gained through leveraging M2M to solve business challenges 是一个定语从句，修饰和限定 the benefits。在该从句中，through leveraging M2M 是介词短语，作方式状语，to solve business challenges 是动词不定式短语，作目的状语。这两个状语都修饰谓语 are gained。

[3] Devices in remote locations where it's difficult to run cables can also be attached to a wireless network quickly and easily.

本句中，in remote locations where it's difficult to run cables 是介词短语，作定语，修饰和限定 Devices。其中 where it's difficult to run cables 是一个定语从句，修饰和限定 remote locations。

[4] Mobile devices can connect into a network, which is not possible with a fixed line network, allowing data communication with all devices.

本句中，which is not possible with a fixed line network 是一个非限定性定语从句，对主句进行补充说明，which 指整个主句。allowing data communication with all devices 是一个现在分词短语，作结果状语。

[5] Wireless networks deliver real time information to mobile devices enabling information to be received or sent whenever and wherever it is required, so organizations have access to live information for effective and fast decisionmaking.

本句中，enabling information to be received or sent whenever and wherever it is required 是一个现在分词短语，作定语，修饰和限定 mobile devices。so organizations have access to live information for effective and fast decision making 是一个目的状语从句，修饰主句的谓语 deliver。

Exercises

【EX.1】Answer the following questions according to the text.

1. What does M2M refer to?
2. What is the characteristic of M2M systems? What does it mean?
3. What is wireless M2M?
4. How many fields can M2M be used in the text?
5. Can M2M offer centralized service support and data management?

6. Are devices that are wirelessly connected to a network limited to a physical location?
7. What do wireless networks deliver to mobile devices?
8. Which one can be deployed more quickly, a wireless network or a fixed one? Why?
9. Which one is less cost effective, CDMA cellular or PSTN/fixed line?
10. How is CDMA airtime/data usage billed?

【EX.2】 Translate the following terms or phrases from English into Chinese and vice versa.

1.	centralize	1.	
2.	modem	2.	
3.	mobility	3.	
4.	server	4.	
5.	logistic	5.	
6.	ensure	6.	
7.	automation	7.	
8.	remote access control	8.	
9.	building automation	9.	
10.	product lifecycle	10.	
11.	移动设备	11.	
12.	*n.* 功能，作用	12.	
13.	*n.* 信息通信业务，远程信息处理	13.	
14.	*n.* 停工期	14.	
15.	*v.* 部署，展开，配置	15.	

【EX.3】 Translate the following sentences into Chinese.
1. Developments in electronic, sonic, and laser technology will provide telemetry that will increase accuracy and timely management response.
2. Have VLAN function is an important indicator to measure LAN switches.
3. Wireless area network (WLAN) is an important part in computer networks.
4. Automation of the factory has greatly increased its productivity.
5. What is the difference between logistics and supply chain management?
6. Logic simulation is an important component of automatic design of digital circuits.
7. In database, a field contains information about an entity.
8. You can then distribute your custom components, or you can move them from test to production.
9. With a 10 Mbps cable modem, that same file can be downloaded in 8 seconds.
10. As our technological powers grow, the portability and flexibility of our computer hardware grows, too.

【EX.4】 Complete the following passage with appropriate words in the box.

waste	services	adapted to	changed	public
security	free	transportation	optimize	urban

Statistics, forecasts and population studies confirm the continuous migration of population towards cities. There, people find jobs, better access to __1__ and better living conditions. The current __2__ environment is not __3__ this massive migration. This means, new challenges in the fields of __4__, environmental issues, __5__ systems, water distribution and — more general — resource management will rapidly occur.

Today, large cities are faced with problems such as __6__ or misuse of resources, which could be __7__ by increased in-time information. In digital cities, people will arrive just in time for their __8__ transportation as exact information is provided to their device in due time. Even parking your car will be easier as __9__ parking spots around you are shown on your device.

Real time information that is always available will __10__ time, save energy and make life easier.

【EX.5】 Translate the following passage into Chinese.

Digital City

The term digital city or digital community (smart community, information city and e-city are also used) refers to a connected community that combines broadband communications infrastructure; flexible, service-oriented computing infrastructure based on open industry standards; and innovative services to meet the needs of governments and their employees, citizens and businesses. The geographical dimension (space) of digital communities vary: they can be extended from a city district up to a multi-million metropolis.

While wireless infrastructure is a key element of digital city infrastructure, it is only a first step. The digital city may require hard-wired broadband infrastructure, and it is much more than just the network. A digital city provides interoperable, Internet-based government services that enable ubiquitous connectivity to transform key government processes, both internally across departments and employees and externally to citizens and businesses. digital city services are accessible through wireless mobile devices and are enabled by services oriented enterprise architecture including Web services, the extensible markup language (XML), and mobilized software applications.

扫码听音频

Text B

Types of Transmission Media

Transmission media can be defined as a means to setup a communication pathway in order to convey the information between the sender and receiver in the form of electromagnetic signal waves. It is operated using various physical elements, so it is placed underneath the physical layer. The material used for transmission is either copper-based or fibre-based to transmit electric or light signals respectively.

The transmission phenomenon can be explained in layman terms as it is an objective conduit between two physical elements, namely the transmitter and the receiver. This trail lane is used for sending and receiving various signals, depending on the material used for connecting the transmitters and receivers, that is, the transmission media.

The transmission media are chiefly categorized as guided media and unguided media[1], which can be further classified in accordance to the type and quality of the transmission.

1. Guided Media

Guided media is a type of transmission media that can be otherwise known as wired transmission. It is also termed as bounded transmission media, as it is bound to a specific limit in the communication network. In this guided media, the transmission signal properties are restricted and focused in a fixed constricted channel, which can be implemented with the help of bodily wired contacts. One of the notable properties of the guided media is the velocity of transmission, which is observed to be at high speed. Other reasons that make the users choose guided media over unguided media are the security provided in transmission and the coverage of the network to be controlled inside a smaller geographical area.

The guided media transmission is further classified into five different types based on the type of connecting material used for creating the network.

1.1 Twisted Pair Cable

The twisted pair cable can be defined as a cable formed by twisting two different shielded cables[2] around each other to form a single cable. The shields are usually made of insulated materials that allow both cables to transmit on their own. This twisted cable is then placed inside a protective layer around it for the sake of ease of use.

These twisted pair cables are available in two different forms, where one is shielded, and another is unshielded.

1.2 Shielded Twisted Pair Cable

Shielded cables are nothing but the transmission media that has exceptional casing to obstruct any or all the peripheral intrusions during the transmission process. These cables are known for their high performance that doesn't allow signal crossings and faster transmission rates. A typical application of the shielded twisted pair cable is the telephone line seen in

---

1 Unguided media transport electromagnetic waves without using a physical conductor. It is also known as unbounded or wireless media, and does not rely on physical pathways to transmit signals. Instead, they use wireless communication methods to propagate signals through the air or free space.

2 A shielded cable or screened cable is an electrical cable that has a common conductive layer around its conductors for electromagnetic shielding.

domestic utilities. Like any other media, shielded twisted pair cables have their own cons in them. They are more expensive than other cables, difficult to install, and they require a huge volume of wires.

1.3 Unshielded Twisted Pair Cable

This type of cable doesn't have the casing, as the name says, and has many qualities inversely proportional to the shielded cable type. These cables are less expensive, easier to install and with faster transmitting abilities. However, it invites outer interference, which leads to lesser performance qualities.

1.4 Optical Fibre Cable[1]

Optical fibre cables can be defined as the cables made of glass material, which uses the light signals for transmission purposes. The reflection principles are used for light signal transmission through the cables. It is known for letting bulky data to be transmitted with higher bandwidth and lesser electromagnetic interference[2] during transmission. Since the material is not corrosive in nature and is very light, these cables are preferred over twisted cables in most cases. However, it has some disadvantages.It is difficult to install and maintain.It is more expensive than other types of transmission media.

1.5 Coaxial Cable

Coaxial cables are made of plastic layering on the outside and two conducting materials placed in parallel to one another while being wrapped in individual insulating layers around them. They are used for transmitting data with dedicated cables or a single cable cracked into different bandwidths, and are referred to as baseband mode and broadband mode, respectively. A well-known application of this type of cable is for providing television network in the houses. A few of the advantageous qualities of this type of cable are exceptional bandwidth range, simple installation or maintenance, and not as expensive as other cable types. The coaxial cable can form a single cable network, and if it fails, the network is disordered completely.

2. Unguided Media

As the name says, unguided media is not a guided media, which simply means that the network created using this type of transmission media cannot be bound to a certain physical plan. It can be defined as a wireless transmission media with no physical medium to provide the connection to the nodes or servers in the network. The electromagnetic signal waves are transmitted in the air across a larger geographical area, and so it is less secure than the guided media.

---

1 A fibre optic cable is a network cable that contains strands of glass fibres inside an insulated casing. They're designed for long-distance, high-performance data networking, and telecommunications. Compared to wired cables, fibre optic cables provide higher bandwidth and transmit data over longer distances. Fibre optic cables support much of the world's internet, cable television, and telephone systems.

2 Electromagnetic interference (EMI) is unwanted noise or interference in an electrical path or circuit caused by an outside source. It is also known as radio frequency interference. EMI can cause electronics to operate poorly, malfunction or stop working completely.

This type of transmission media is further classified into three types with respect to the signals used for the transmission.

2.1 Radio Waves

Radio waves are the simplest form of transmission signal, which do not involve any complicated steps to create and transmit. This signal generally ranges between 3 kHz and 1 GHz of frequency, and the signal types can be of AM and FM signals. The main applications of this transmission media are the cordless phones for domestic or official use, as well as radio devices used as communication elements for mass media. These radio waves can be used for ground-based or satellite communications.

2.2 Micro Waves

Micro waves are the type of transmission media that uses antennas as the main element for sending and receiving the data. The area coverage provided by these signals is directly related to the elevation of the antenna placement. The signal range for this type of transmission is between 1 GHz and 300 GHz, which are usually used for mobile phone and television networks.

2.3 Infrared

Infrared is another way of transmitting the data inside a small area, which cannot pass through the obstacles and doesn't give in for interference. These waves come in a range of 300 GHz to 400 THz and can be used for wireless peripheral devices like mouse, remotes, keyboards, printers, etc.

Transmission media are the essential constituents to setup a flawless network, which can operate on its own without any glitches in sending and receiving the data across the network. Without a medium to transmit the contents inside a network, the network setup cannot be complete or become a completely functioning system.

New Words

setup	['setʌp]	n. 建立
convey	[kən'veɪ]	vt. 传输，传递
sender	['sendə]	n. 发送器；发送者
receiver	[rɪ'si:və]	n. 接收器；接受者
signal	['sɪgnəl]	n. 信号
		v. 发信号
wave	[weɪv]	n. 波
copper-based	['kɒpə beɪst]	adj. 铜基的
fibre-based	['faɪbə beɪst]	adj. 光纤基的
phenomenon	[fə'nɒmɪnən]	n. 现象，事件
layman	['leɪmən]	n. 门外汉，外行
conduit	['kɒndjuɪt]	n. [电]导管；渠道
transmitter	[træns'mɪtə]	n. 发射器，发射机

bound	[baʊnd]	vt. 给……划界，限制
		n. 界限，限制
channel	['tʃænl]	n. 通道，频道
notable	['nəʊtəbl]	adj. 显著的，著名的
velocity	[və'lɒsəti]	n. 速率，速度
coverage	['kʌvərɪdʒ]	n. 范围，规模
shield	[ʃi:ld]	vt. 屏蔽
unshield	[ʌnʃi:ld]	vt. 非屏蔽
exceptional	[ɪk'sepʃənl]	adj. 独特的
obstruct	[əb'strʌkt]	vt. 阻止；阻碍
domestic	[də'mestɪk]	adj. 家庭的，家的
interference	[ˌɪntə'fɪərəns]	n. 干涉，干扰
bulky	['bʌlki]	adj. 庞大的
corrosive	[kə'rəʊsɪv]	adj. 腐蚀性的；侵蚀性的
		n. 腐蚀性物品
plastic	['plæstɪk]	n. 塑料制品
		adj. 塑料的
baseband	['beɪsbænd]	n. 基带
broadband	['brɔ:dbænd]	n. 宽带
disordered	[dɪs'ɔ:dəd]	adj. 混乱的，杂乱的
node	[nəʊd]	n. 节点
ground-based	['graʊnd 'beɪst]	adj. 陆基的，以地面为基础的
antenna	[æn'tenə]	n. 天线
elevation	[ˌelɪ'veɪʃn]	n. 高度
infrared	[ˌɪnfrə'red]	adj. 红外线的
		n. 红外线
obstacle	['ɒbstəkl]	n. 障碍（物）
remote	[rɪ'məʊt]	n. 遥控器
		adj. 远程的
keyboard	['ki:bɔ:d]	n. 键盘
printer	['prɪntə]	n. 打印机，印刷机
constituent	[kən'stɪtjuənt]	n. 成分，构成部分
		adj. 构成的，组成的
glitch	[glɪtʃ]	n. 小过失，差错

Phrases

communication pathway	通信通道
guided media	导引性介质

unguided media	非导引性介质
in accordance to	依照，根据
wired transmission	有线传输
geographical area	地理区域
twisted pair cable	双绞线电缆
shielded cable	屏蔽电缆
single cable	单芯电缆
insulated material	绝缘材料
protective layer	保护层
telephone line	电话线
optical fibre cable	光纤电缆，光缆
reflection principle	反射原理
electromagnetic interference	电磁干扰
coaxial cable	同轴电缆
conducting material	导电材料
be wrapped in	被包裹在
insulating layer	绝缘层
dedicated cable	专用电缆
television network	电视网络
be defined as	被定义为
wireless transmission media	无线传输介质
radio wave	无线电波
radio device	无线电设备
satellite communication	卫星通信
micro wave	微波
signal range	信号范围
mobile phone	移动电话
wireless peripheral device	无线外围设备，无线外部设备

Abbreviations

kHz (kiloHertz)	千赫兹
GHz (GigaHertz)	千兆赫兹
AM (Amplitude Modulation)	调幅
FM (Frequency Modulation)	调频

Exercises

【EX.6】 Answer the following questions according to the text.

1. What can transmission media be defined as?

2. What are transmission media chiefly categorized as?
3. What is guided media? What is it also termed as and why?
4. What can the twisted pair cable be defined as?
5. What are shielded cables known for?
6. What can optical fibre cables be defined as?
7. What are coaxial cables made of ?
8. What can unguided media be defined as?
9. What are the main applications of radio waves?
10. What are micro waves?

【EX.7】Translate the following terms or phrases from English into Chinese and vice versa.

1.	antenna	1.	
2.	baseband	2.	
3.	broadband	3.	
4.	channel	4.	
5.	insulating layer	5.	
6.	infrared	6.	
7.	interference	7.	
8.	communication pathway	8.	
9.	electromagnetic interference	9.	
10.	insulated material	10.	
11.	光纤电缆，光缆	11.	
12.	无线电波	12.	
13.	无线传输介质	13.	
14.	*n.* 接收器；接受者	14.	
15.	*n.* 信号 *v.* 发信号	15.	

【EX.8】Translate the following sentences into Chinese.

1. With a wireless setup, you stick battery-powered sensors up around your home that keep an eye on windows, doors, motion, and more.
2. To receive an encrypted message, you must make sure that the sender has your public key in advance.
3. You have pressed the receiver button.
4. The signal will be converted into digital code.
5. Existing copper-based switches can be used further in the network concept.
6. The transmitters will send a signal automatically to a local base station.
7. They have been accused of deliberately causing interference to transmissions.

8. The digital baseband signal's transmission is one of the important components of digital communication systems.
9. Internet multiplayer games are responsible for much of the increase in broadband use.
10. The radio may have an antenna attached to it.

Reading Material

IoT Networks

An IoT network is a group of hardware (including sensors, gadgets[1], appliances) and software that interact with one another and share data and information without the need for human interaction. Businesses may now gather new insights from devices through IoT networks thanks to cloud and edge computing capabilities. Organizations may now monitor environmental, geospatial[2], and atmospheric[3] variables in real time because of this bridging of the digital and physical worlds. Businesses can quickly respond to environmental changes when combined with automation, resulting in less downtime, more significant insights, and increased productivity.

1. How Does an IoT Network Work?

IoT networks use small, low-cost[4] sensors to gather data about the surroundings. For instance, farmers employ IoT sensors to track moisture levels, while industrial facilities utilize the same sensors to track pipe pressure. IoT sensors provide a wide range of configuration options and can track hundreds of distinct changes.

IoT sensors feed information back[5] to the cloud or an edge computing device constantly for processing. Instead of sending vast data streams, IoT devices usually utilize less power and provide smaller quantities of data. Since edge computing reduces the distance between the sensor and the server, it is frequently chosen by businesses that need the lowest latency and quickest reaction time[6]. Businesses may select from a variety of IoT networks depending on the technology and use case to achieve their objectives. WiFi or cellular connections are the two methods through which sensors often convey[7] their data.

The software then analyses and stores the data in the cloud or on an edge server after it has been collected. Several systems employ artificial intelligence and machine learning to take actions when particular data is transmitted from a sensor. Businesses combine automation and IoT networks to coordinate device management in a low-cost, predictable, and scalable way. Enterprises are able to monitor anything from machine maintenance to the weather outdoors because IoT management solutions can handle data from diverse platforms.

---

1 gadget ['gædʒɪt] *n.* 小装置
2 geospatial [ˌdʒiːəʊ'speɪʃ(ə)l] *adj.* 地理空间的
3 atmospheric [ˌætməs'ferɪk] *adj.* 大气的
4 low-cost [ˌləʊ'kɒst] *adj.* 价格便宜的，廉价的
5 feed information back：反馈信息
6 reaction time：反应时间
7 convey [kən'veɪ] *vt.* 传达，传递

2. Types of IoT Networks

2.1 Cellular

IoT devices may interact using cellular networks, the same mobile networks used by smartphones. These networks are not always thought to be the most excellent[1] option for IoT devices because they were first created for power-hungry[2] gadgets like smartphones. Later, the cellular sector created new technologies that were more suited for IoT use cases. At present this kind of wireless network is widely used and regarded as a dependable[3] and secure form of IoT communication.

2.2 WiFi

WiFi is a standard option for IoT networks since many companies already have WiFi coverage across their infrastructure. For stationary[4] IoT sensors that must communicate data over a medium distance, WiFi is a reliable solution. To assist and improve the reliability of their sensors, WiFi administrators could segment[5] IoT sensors on a distinct subnet. WiFi IoT networks do have certain disadvantages, though. WiFi networks don't have as much coverage as cellular networks because of their power restrictions[6]. Mobile IoT sensors may have connection problems on WiFi networks since WiFi networks don't handle device handover[7] as efficiently as as cellular networks do.

2.3 Local Area Network (LAN) and Personal Area Network (PAN)

Local area network and personal area network are networks that only span[8] relatively limited distances. Although data transport via LAN and PAN networks is often thought to be cost-effective, it is not always dependable. WiFi and Bluetooth are two wireless personal and local area network technologies that are often used in IoT connectivity solutions. When numerous access points are included in a more extensive network, WiFi may be utilized for dispersed applications in addition to local ones. A single battery powered by bluetooth low energy (BLE)[9] might last up to five years if the device is not continually receiving data. BLE is a more energy-efficient wireless network protocol.

2.4 Low Power Wide Area Networks (LPWAN)

IoT devices that use LPWANs transmit little data packets, rarely over great distances. This kind of wireless network was created in response to the early difficulties with cellular communication. LPWAN is marketed as having a more excellent range than WiFi and Bluetooth

---

1 excellent ['eksələnt] *adj.* 卓越的，杰出的，优秀的
2 power-hungry ['paʊər 'hʌŋgri] *adj.* 功耗大的
3 dependable [dɪ'pendəbl] *adj.* 可信赖的，可靠的
4 stationary ['steɪʃənri] *adj.* 固定的；静止的
5 segment ['segmənt] *v.* 分割，划分
6 restriction [rɪ'strɪkʃn] *n.* 限制，限定
7 handover ['hændəʊvə] *n.* 切换；移交，交接
8 span [spæn] *vt.* 跨越时间或空间
9 bluetooth low energy：低功耗蓝牙

while consuming less power than cellular. LoRaWAN, which operates on the LoRa (long-range) communication network, is a well-known and widely used IoT network protocol in this category. LoRaWAN has benefits for IoT devices, including reduced power consumption (for longer battery life[1]) and relatively affordable chipsets[2]. A single base station or gateway operating on a long-range network is capable of delivering service to a very vast area—a few kilometers in congested[3] metropolitan areas—under the right circumstances.

2.5 Mesh[4] Networks

The connection configuration of mesh networks—how the parts communicate with one another—is the most effective way to characterize[5] them. In mesh networks, all sensor nodes work together to share data among themselves so that it may reach the gateway. One illustration of an IoT wireless network technology is Zigbee. Mesh networks have a relatively limited range, so you might need to add more sensors throughout a building or utilize repeaters to achieve the coverage you need for your application. Additionally, the way how these networks interact can lead to excessive power consumption, particularly if you want fast communications, as in the case of an application for intelligent lighting. Mesh networks are a standard option since they are also very resilient[6], adept at locating the data transmission pathways that are both quick and reliable, and simple to set up.

3. Conclusion

The five types of IoT networks mentioned above are an absolute fit for most businesses looking for solutions to their problems in the Internet of Things. These networks are a combination of wired and wireless networks for IoT-connected devices.

参考译文

Text A 什么是 M2M？

M2M 指两个或更多的机器之间的数据通信。M2M 普遍被翻译为机器对机器，有时也译为人对机器、机器对人等。蜂窝电话服务提供商使用公共无线网络实现 M2M、遥测和远程信息处理。

通常，M2M 系统执行具体的任务，这意味着一个给定的系统为一个特定的设备而建立，或为一个行业中非常有限的一类设备而建立。这只是 M2M 市场的一个特点，该市场尚在起步阶段，还没有形成统一的标准。功能是重复的——每个专用系统都重复类似系统已经实现的许多功能。

---

1 battery life：电池寿命
2 chipset ['tʃɪpset] n. 芯片集，芯片组
3 congest [kən'dʒest] vt. 拥挤
4 mesh [meʃ] n. 网状物
5 characterize ['kærəktəraɪz] vt. 描述……的特性，具有……的特征
6 resilient [rɪ'zɪlɪənt] adj. 有弹性的；能复原的

无线 M2M 是通过无线技术（如 CDMA）的机器对机器通信。

1. 何处可用 M2M？

M2M 使各个行业中的众多个体、公司、社区和组织受益。下面是一些利用 M2M 解决方案获得更多好处的"新兴领域"：

- 电信——IP 网络互连/无线广域网/移动学习。
- 制造——供应链/库存管理/工厂自动化。
- 轻工业——传感器监测/远程访问控制/设施。
- 运输——资产跟踪/物流管理。
- 零售——销售点/亭子/数字标牌。
- 信息技术——零件市场/车载解决方案。
- 电子商务/移动商务解决方案。
- 不动产——楼宇自动化。
- 安全——视频监控。

2. M2M 有何益处？

以下是通过 M2M 解决业务挑战可得到的一些好处：

- 成本——效益的预防性维护和服务质量。
- 通过外包的故障排除实现快速响应。
- 集中服务支持和数据管理。
- 在整个产品生命周期不断获得收入。
- 通过使停机时间最短来增加收入。
- 远程诊断。

3. M2M 解决方案由什么组成？

通常，M2M 解决方案是由以下几部分构成：

- 收集数据和/或监测变化的装置或传感器。
- 发送和接收数据的应用程序和数据库。
- 发送和/或接收传感器数据的中央服务器。
- （用固定线路或无线）连接到一个中央服务器的装置或传感器。
- 在设备、传感器和中央服务器设备之间交换数据的调制解调器。
- 保证数据传输安全性及监控和管理设备或传感器网络连通性的应用程序。

4. 无线 M2M 有何益处？

4.1 灵活性

以无线方式连接网络的设备不受物理位置的限制。当需要时可以灵活地移动这些设备。例如，如果一个自动售货机处客流不足，只需搬到其他地方。在偏远的难以布置电缆的地方，可以快速和容易地把设备连接到无线网络。

4.2 移动性

移动设备可以连接不能用固定线路连接的网络，实现了所有设备的数据通信。

4.3 访问信息

无线网络给移动设备提供实时信息，使得这些设备能够随时按需接收或发送信息，这样组织就可以访问即时信息，以便有效和快速地决策。

4.4 独立网络

无线网络可以快速方便地部署到一个建筑内或不需要与现有的固定线路网络整合的地方，从而提供一个独立的安全网络。

4.5 速度

由于无须在设备之间布线，因此无线网络的部署比有线网络快得多。这样能够更快地使用设备，并节约成本。

4.6 成本

节约成本与部署速度、消除电缆成本和降低通信成本相关，像 CDMA 蜂窝网比 PSTN/固定线路更划算。

4.7 无线 M2M 机器如何通信？

M2M 无线解决方案通过无线调制解调器使用即通的 CDMA 蜂窝网通信，可以实现数据的即时通信，速度更快。CDMA 通话/数据将根据所使用的数据量而不是根据持续连接的时间收费。

Unit 3

Text A

IoT Protocols

1. MQTT

MQTT is a widely adopted security protocol in the realm of IoT security. MQTT, which stands for message queuing telemetry transport, is a client-server communication messaging transport protocol. It operates over TCP/IP or other protocols that offer reliable, lossless, and bidirectional connections.

Features of MQTT:

MQTT is a lightweight and straightforward protocol that facilitates rapid and efficient data transmission.It is specifically designed for use with constrained devices and networks that have low bandwidth, high latency, or unreliability.

The protocol's minimal use of data packets[1] results in reduced network usage, while its optimal power consumption helps to conserve the battery life of connected devices, making it an ideal choice for mobile phones and wearables.

MQTT is based on messaging techniques, which ensures fast and reliable communication. As such, it is well-suited for use in IoT applications.

Where is it used?

The security of MQTT is structured into distinct layers, namely the network, transport, and application levels, each of which serves to thwart a particular form of attack. Given that MQTT is a protocol that is lightweight in nature, it incorporates only a limited number of security mechanisms. To bolster security, MQTT implementations frequently leverage other security

---

1　A data packet is a unit of data made into a single package that travels along a given network path. Data packets are used in internet protocol (IP) transmissions for data that navigates the Web, and in other kinds of networks.

standards such as SSL[1]/TLS[2] for transport encryption, VPN at the network level to ensure a physically secure network, and the use of username/password. In addition, a client identifier is transmitted with data packets to authenticate devices at the application level.

2. CoAP

The constraint application protocol (CoAP) is a Web transfer protocol that has been specifically designed to cater to the requirements of constrained devices, such as microcontrollers, and the low power or lossy networks that they operate on. It is widely recognized as one of the most popular protocols for securing IoT applications.

Features of CoAP:

Like HTTP, it is founded on the REST architecture.

Clients utilize methods such as GET, PUT, POST, and DELETE to access resources provided by servers through URLs.

CoAP is specifically engineered to operate on microcontrollers. It is an ideal protocol for the internet of things, which necessitates millions of low-cost nodes.

CoAP is resource-efficient, which requires minimal resources on both the device and the network. It employs UDP on IP instead of a complex transport stack.

Where is it used?

The CoAP utilizes the UDP for information transportation and consequently depends on the security aspects of UDP to safeguard the information. CoAP employs datagram transport layer security (DTLS) over UDP for enhanced security.

CoAP has been developed with a straightforward and user-friendly interface that seamlessly integrates with HTTP for Web integration. It also offers features such as multicast support and low overhead concerns, thereby contributing to the security of the Internet of Things.

3. DTLS

The datagram transport layer security (DTLS) protocol is a security measure specifically developed for IoT to safeguard data communication between applications that rely on datagrams. DTLS is built on the foundation of the transport layer security (TLS) protocol and it offers an equivalent level of security.

The primary aim of DTLS is to address challenges such as data loss and reordering by making minor modifications to TLS. The DTLS protocol preserves the semantics of the underlying transport layer, thereby avoiding any delays caused by associated stream protocols. However, the application must handle issues such as datagram loss, packet reordering, and data exceeding the size of a datagram network packet.

---

1 Secure sockets layer (SSL) is a security protocol that provides privacy, authentication, and integrity to Internet communications. SSL eventually evolved into transport layer security (TLS).

2 Transport layer security (TLS) is the most widely used security protocol for communications over the Internet. TLS provides three main functions: authentication, encryption, and verification. It encrypts transmissions using a system of certificates and keys, verifies both network entities are authorized to transmit/receive data, and ensures the data hasn't been corrupted.

Features of DTLS:

DTLS employs a retransmission timer to address the challenge of packet loss.

In the event that the timer expires prior to the client receiving the confirmation message from the server, the client will retransmit the data. To mitigate the issue of reordering, each message is assigned a unique sequence number, enabling the determination of whether the subsequent message received is in sequence or not. If it is out of sequence, it is placed in a queue and processed when the sequence number is attained.

Where is it used?

DTLS is commonly used in various applications, including live video feeds, video streaming, gaming, VoIP, and instant messaging. This protocol is particularly suitable for scenarios where low latency is of greater significance than data loss.

4. 6LoWPAN

The 6LoWPAN protocol, which stands for IPv6 over low power wireless personal area networks, is specifically designed for low-power networks such as wireless sensor networks and IoT systems.

Features of 6LoWPAN:

6LoWPAN is a protocol utilized for transmitting data packets in the form of IPv6 across diverse networks.It offers end-to-end IPv6 connectivity, thereby enabling direct access to a broad range of networks, including the Internet. Additionally, 6LoWPAN is employed to safeguard communications between end-users and sensor networks.

To ensure security in the IoT, 6LoWPAN utilizes AES-128 link layer security, as defined in IEEE 802.15.4. Link authentication and encryption are utilized to provide security, and further security is provided to transport layer security mechanisms that operate over TCP.

Where is it used?

6LoWPAN is a pivotal technology in various domains such as smart home automation, industrial monitoring, smart grids, and general automation.

5. ZigBee

ZigBee is widely regarded as a cutting-edge protocol that offers robust security for IoT devices and applications. This technology facilitates seamless machine-to-machine communication over distances ranging from 10 to 100 meters, making it ideal for low-powered embedded devices such as radio systems. Additionally, ZigBee is an open-source wireless technology that is both cost-effective and highly efficient.

Features of ZigBee:

ZigBee offers standardization across all layers, promoting compatibility among products from various manufacturers.Its mesh architecture facilitates connectivity with nearby devices, thereby expanding the network and enhancing its flexibility.

The implementation of "Green Power" by ZigBee results in reduced energy consumption and cost. Additionally, ZigBee supports a high number of devices, approximately 6550, contributing to the scalability of networks.

Where is it used?

ZigBee is mainly used in home automation, medical data collection, industrial control systems, meter reading system, light control system, commercial, government markets worldwide, home networking, etc.

6. AMQP

AMPQ is a highly efficient, portable, and multichannel messaging protocol that prioritizes security. The protocol offers authentication and encryption through SASL[1] or TLS, which rely on a transport protocol like TCP.

Features of AMQP:

The AMQP protocol is developed with the aim of facilitating communication between a diverse range of applications and systems, regardless of their internal architecture. This has resulted in the standardization of business communications on an industrial scale.

Where is it used?

The protocol is utilized in client/server communication as well as in the management of IoT devices. AMPQ boasts of its efficiency, portability, multichannel capabilities, and security features.

7. DDS

DDS is a publish-subscribe protocol that differs from MQTT in that it does not require a server connection. Instead, DDS utilizes a brokerless architecture, resulting in a high-speed and high-performance protocol that is not reliant on any intermediary system. Developed by the object management group (OMG), DDS is specifically designed for device-to-device communications.

Features of DDS:

The DDS technology facilitates the creation of open architecture systems that are modular and loosely coupled.

It achieves this by enabling well-defined interfaces between subsystems and components, thereby eliminating the closed and proprietary architecture.

This approach reduces the costs associated with integration, maintenance, and upgrades, while promoting competition and ease of reuse at the middleware and subsystem levels.

Moreover, DDS standardizes messaging semantics, which enhances the system's robustness and reduces the overall development and integration costs.

Where is it used?

---

1 Simple authentication and security layer (SASL) is a framework for authentication and data security in Internet protocols. It decouples authentication mechanisms from application protocols, in theory allowing any authentication mechanism supported by SASL to be used in any application protocol that uses SASL.

DDS caters to the real-time data exchange requirements of various applications in the aerospace, defense, air-traffic control, autonomous vehicles, medical devices, robotics, simulation and testing, smart grid management, transportation systems, and other related domains.

New Words

adopt	[ə'dɒpt]	vt. 采用，采取
realm	[relm]	n. 领域，范围
operate	['ɒpəreɪt]	v. 运行；操作
reliable	[rɪ'laɪəbl]	adj. 可靠的；可信赖的
lossless	['lɒsləs]	adj. 无损的；无损耗的
feature	['fiːtʃə]	n. 特征，特点
lightweight	['laɪtweɪt]	adj. 轻量的
straightforward	[ˌstreɪt'fɔːwəd]	adj. 明确的
facilitate	[fə'sɪlɪteɪt]	vt. 促进；使容易；帮助
bandwidth	['bændwɪdθ]	n. 带宽
latency	['leɪtənsɪ]	n. 延迟
unreliability	[ˌʌnrɪˌlaɪə'bɪlətɪ]	n. 不安全性，不可靠性
wearable	['weərəbl]	adj. 可穿用的，可佩戴的
transport	['trænspɔːt]	vt. 传输，运输
attack	[ə'tæk]	v. & n. 攻击，进攻
incorporate	[ɪn'kɔːpəreɪt]	vi. 合并；混合
bolster	['bəʊlstə]	vt. 支持，支撑
username	['juːzəneɪm]	n. 用户名
password	['pɑːswɜːd]	n. 密码；口令
client	['klaɪənt]	n. 客户，客户端
identifier	[aɪ'dentɪfaɪə]	n. 标识符
authenticate	[ɔː'θentɪkeɪt]	vt. 验证
recognize	['rekəgnaɪz]	vt. 识别；承认
safeguard	['seɪfgɑːd]	n. & vt. 保护，保卫；防护
enhance	[ɪn'hɑːns]	vt. 增强，加强；提高
seamlessly	['siːmləslɪ]	adv. 无缝地
integrate	['ɪntɪgreɪt]	v. 集成，合并；成为一体
equivalent	[ɪ'kwɪvələnt]	adj. 相等的，相当的，等效的；等价的 n. 对等物
modification	[ˌmɒdɪfɪ'keɪʃn]	n. 修改，修正，变更
delay	[dɪ'leɪ]	n. 延迟
stream	[striːm]	n. 流
timer	['taɪmə]	n. 定时器，计时器

confirmation	[ˌkɒnfə'meɪʃn]	n.	确认；证实
determination	[dɪˌtɜːmɪ'neɪʃn]	n.	确定
subsequent	['sʌbsɪkwənt]	adj.	后来的；随后的
attain	[ə'teɪn]	vi.	获得；达到
wireless	['waɪələs]	adj.	无线的
end-to-end	[endtuːend]	adj.	端到端的
pivotal	['pɪvətl]	adj.	关键的；中枢的
general	['dʒenrəl]	adj.	通用的，普遍的
cutting-edge	['kʌtɪŋ 'edʒ]	adj.	前沿的
robust	[rəʊ'bʌst]	adj.	强健的；坚固的，结实的
radio	['reɪdɪəʊ]	n.	无线电
open-source	['əʊpən sɔːs]	adj.	开源的，提供源程序的
standardization	[ˌstændədaɪ'zeɪʃn]	n.	标准化；规范化
portable	['pɔːtəbl]	adj.	轻便的；手提的
multichannel	['mʌltɪtʃænl]	adj.	多通道；多通路；多波段的
brokerless	['brəʊkələs]	adj.	无代理的
intermediary	[ˌɪntə'miːdɪəri]	adj.	中间的
subsystem	['sʌb'sɪstəm]	n.	子系统，分系统
simulation	[ˌsɪmju'leɪʃn]	n.	模拟
robotic	[rəʊ'bɒtɪk]	adj.	机器人的；自动的

Phrases

bidirectional connection	双向连接
data transmission	数据传输
constrained device	受限设备，受限装置
data packet	数据包
power consumption	能量消耗；耗电量
be structured into	被组织成，被构造为
security mechanism	安全机制
application level	应用层
low-cost node	低成本节点
user-friendly interface	用户友好界面
unique sequence number	唯一序列号
instant messaging	即时通信（服务），即时信息传输
be suitable for ...	适合……的
link layer	链路层
industrial monitoring	工业监测
smart grid	智能电网

be regarded as	被认为
industrial control system	工业控制系统
home networking	家庭联网
client/server communication	客户端/服务器通信
open architecture system	开放体系系统
air-traffic control	空中交通管制
autonomous vehicle	自动驾驶车辆
medical device	医疗设备
transportation system	运输系统

Abbreviations

MQTT (Message Queuing Telemetry Transport)	消息队列遥测传输
TCP/IP (Transmission Control Protocol/Internet Protocol)	传输控制协议/网际协议
SSL (Secure Socket Layer)	安全套接字层
TLS (Transport Layer Security)	传输层安全协议
VNN (Virtual Native Network)	虚拟专用网
CoAP (Constraint Application Protocol)	约束应用程序协议
HTTP (Hypertext Transfer Protocol)	超文本传输协议
REST (Representational State Transfer)	描述性状态转移
DTLS (Datagram Transport Layer Security)	数据报传输层安全
VoIP (Voice over Internet Protocol)	互联网电话
6LoWPAN (IPv6 over Low Power Wireless Personal Area Networks)	低功耗无线个人区域网上的 IPv6
IEEE (Institute of Electrical and Electronics Engineers)	电气电子工程师学会
AMQP (Advanced Message Queuing Protocol)	高级消息队列协议
SASL (Simple Authentication and Security Layer)	简单身份验证和安全层
DDS (Data Distribution Service)	数据分发服务
OMG (Object Management Group)	对象管理组

Analysis of Difficult Sentences

[1] MQTT, which stands for message queuing telemetry transport, is a client-server communication messaging transport protocol.

本句中，which stands for message queuing telemetry transport 是一个非限定性定语从句，对 MQTT 进行补充说明。

[2] The security of MQTT is structured into distinct layers, namely the network, transport, and application levels, each of which serves to thwart a particular form of attack.

本句中，namely the network, transport, and application levels 对 distinct layers 进行补充说明。each of which serves to thwart a particular form of attack 是一个非限定性定语从句，也

对 distinct layers 进行补充说明。

英语中，名词/代词/数词+of+which / whom 可以引导一个非限定性定语从句。例如：

Peter bough several books, five of which were on management.

皮特买了一些书，其中 5 本是管理方面的。

I bought a new mobile phone online last week, the price of which was very reasonable.

我上周在网上买了一部新手机，价格很合理。

Our manager has a lot of friends, some of whom are very famous businessmen.

我们经理有许多朋友，其中一些是非常著名的商界人士。

[3] The constraint application protocol (CoAP) is a Web transfer protocol that has been specifically designed to cater to the requirements of constrained devices, such as microcontrollers, and the low power or lossy networks that they operate on.

本句中，that has been specifically designed to cater to the requirements of constrained devices, such as microcontrollers, and the low power or lossy networks that they operate on 是一个定语从句，修饰和限定 a Web transfer protocol。在该从句中，such as microcontrollers 是对 constrained devices 的举例说明。that they operate on 是一个定语从句，修饰和限定 the low power or lossy networks。

[4] The datagram transport layer security (DTLS) protocol is a security measure specifically developed for the IoT to safeguard data communication between applications that rely on datagrams.

本句中，specifically developed for IoT to safeguard data communication between applications that rely on datagrams 是一个过去分词短语，作定语，修饰和限定 a security measure。to safeguard data communication between applications that rely on datagrams 是一个动词不定式短语，作目的状语，修饰 developed。

[5] DDS is a publish-subscribe protocol that differs from MQTT in that it does not require a server connection.

本句中，that differs from MQTT in that it does not require a server connection 是一个定语从句，修饰和限定 a publish-subscribe protocol。在该从句中，in that it does not require a server connection 是一个原因状语从句，修饰谓语 differs from。

Exercises

【EX.1】**Answer the following questions according to the text.**

1. What does MQTT stand for? What is it?
2. What is MQTT specifically designed for?
3. What is CoAP widely recognized as?
4. What is the datagram transport layer security (DTLS) protocol?
5. Where is DTLS commonly used?
6. What is the 6LoWPAN protocol specifically designed for?
7. What is ZigBee widely regarded as?

8. Where is ZigBee used?

9. What is the aim that the AMQP protocol is developed with?

10. What does the DDS technology facilitate?

【EX.2】 Translate the following terms or phrases from English into Chinese and vice versa.

1. attack   1. _____
2. bandwidth   2. _____
3. client   3. _____
4. delay   4. _____
5. enhance   5. _____
6. feature   6. _____
7. identifier   7. _____
8. lossless   8. _____
9. multichannel   9. _____
10. application level   10. _____
11. 双向连接   11. _____
12. 数据包   12. _____
13. 链路层   13. _____
14. 智能电网   14. _____
15. n. 模拟   15. _____

【EX.3】 Translate the following sentences into Chinese.

1. Over time, the technology is diffused and adopted by other countries.

2. In the event of the machine not operating correctly, an error code will appear.

3. They need a reliable method to decrease the failure rate.

4. This model embodies many new features.

5. How to reduce the consumption of network bandwidth is a hot point in P2P network research.

6. It's no wonder that new applications for the IoT are moving ahead fast when almost every new device we buy has a plug on the end of it or a wireless connection to the internet.

7. He conceived of the first truly portable computer in 1968.

8. Most smart home technology and devices are wireless.

9. Power management is one of the most important keys for lower power design of wearable computing system.

10. Each computer on a network must have a unique identifier.

【EX.4】 Complete the following passage with appropriate words in the box.

| transmitter | switch | installed | receive | plugged |
| standardize | convey | entertainment | smart | coded |

A smart home or building is a home or building, usually a new one, that is equipped with special structured wiring to enable occupants to remotely control or program an array of automated home electronic devices by entering a single command. For example, a homeowner on vacation can use a Touchtone phone to arm a home security system, control temperature gauges, __1__ appliances on or off, control lighting, program a home theater or entertainment system, and perform many other tasks.

The field of home automation is expanding rapidly as electronic technologies converge. The home network encompasses communications, __2__, security, convenience, and information systems.

A technology known as powerline carrier systems (PCS) is used to send __3__ signals along a home's existing electric wiring to programmable switches, or outlets. These signals __4__ commands that correspond to "addresses" or locations of specific devices, and that control how and when those devices operate. A PCS __5__, for instance, can send a signal along a home's wiring, and a receiver plugged into any electric outlet in the home could __6__ that signal and operate the appliance to which it is attached.

One common protocol for PCS is known as X10, a signaling technique for remotely controlling any device __7__ into an electrical power line. X10 signals, which involve short radio frequency (RF) bursts that represent digital information, enable communication between transmitters and receivers.

In Europe, technology to equip homes with __8__ devices centers on development of the European Installation Bus, or Instabus. This embedded control protocol for digital communication between smart devices consists of a two-wire bus line that is __9__ along with normal electrical wiring. The Instabus line links all appliances to a decentralized communication system and functions like a telephone line over which appliances can be controlled. The European Installation Bus Association is part of Konnex, an association that aims to __10__ home and building networks in Europe.

【EX.5】 Translate the following passage into Chinese.
What Is a "Smart House"?
A smart house is a house that has highly advanced automatic systems for lighting, temperature control, multi-media, security, window and door operations, and many other functions.

A smart home appears "intelligent" because its computer systems can monitor so many aspects of daily living. For example, the refrigerator may be able to inventory its contents, suggest menus, recommend healthy alternatives, and order groceries. The smart home systems might even take care of cleaning the cat's litter box and watering the plants.

However, smart home technology is real, and it's becoming increasingly sophisticated. Coded signals are sent through the home's wiring to switches and outlets that are programmed to operate appliances and electronic devices in every part of the house. Home automation can be especially useful for elderly and disabled persons who wish to live independently.

Text B

IoT Sensors

Sensors are devices that respond to inputs from the physical world and then take those inputs and display them, transmit them for additional processing, or use them in conjunction with artificial intelligence to make decisions or adjust operating conditions. When applied to the Industrial IoT, data collected from sensors is used to help business owners and managers make intelligent decisions about their operations, and help users use the business' products and services more efficiently.

As the IoT initiative expands, more and more sensors are going to be used to monitor and collect data for analysis and processing.

Sensors are designed to respond to specific types of conditions in the physical world, and then generate a signal (usually electrical) that can represent the magnitude of the condition being monitored. Those conditions may be light, heat, sound, distance, pressure, or some other more specific situation, such as the presence or absence of a gas or liquid.

1. Temperature Sensors

Temperature sensors detect the temperature of the air or a physical object and convert that temperature level into an electrical signal that can accurately reflect the measured temperature. These sensors can monitor the temperature of the soil to help with agricultural output or the temperature of a bearing operating in a critical piece of equipment to sense when it might be overheating or nearing the point of failure.

2. Pressure Sensors

Pressure sensors measure the pressure or force per unit area applied to the sensor and can detect things such as atmospheric pressure, the pressure of a stored gas or liquid in a sealed system such as tank or pressure vessel, or the weight of an object.

3. Motion Sensors

Motion sensors or detectors can sense the movement of a physical object by using any one of several technologies, including passive infrared (PIR)[1], microwave detection, or ultrasonic, which uses sound to detect objects. These sensors not only can be used in security and intrusion detection systems, but also can be used to automate the control of doors, sinks, air conditioning and heating, or other systems.

4. Level Sensors

A level sensor is a device used to determine the liquid level flowing in an open or closed system. Liquid level measurement can be divided into two types: continuous measurement and point level measurement. Continuous liquid level sensors are used to accurately measure liquid levels, but the measurement results are correct. The point level sensor is used to determine whether the liquid level is high or low. It is designed to indicate whether the liquid has reached a specific point in the container. For example, a fuel gauge displays the level of fuel in a vehicle's tank and provides a continuous level reading. Some automobiles have a light that illuminates when the fuel level tank is very close to empty, acting as an alarm that warns the driver that fuel is about to run out completely.

5. Image Sensors

Image sensors function to capture images to be digitally stored for processing. Examples are license plate readers as well as facial recognition[2] systems. Automated production lines can use image sensors to detect issues with quality such as how well a surface is painted after leaving the spray booth.

6. Proximity Sensors

Proximity sensors can detect whether an object is approaching the sensor through various technologies.

These technologies include:
- Inductive technologies, which are useful for the detection of metal objects.
- Capacitive technologies, which function on the basis of objects having a different dielectric constant than that of air.
- Photoelectric technologies, which rely on a beam of light to illuminate and reflect back from an object, or
- Ultrasonic technologies, which use a sound signal to detect an object nearing the sensor.

7. Water Quality Sensors

The importance of water to human beings on earth is not only for drinking but as a key ingredient needed in many production processes. So it is necessary to sense and measure water

---

1　A passive infrared sensor (PIR sensor) is an electronic sensor that measures infrared (IR) light radiating from objects in its field of view. They are most often used in PIR-based motion detectors. PIR sensors are commonly used in security alarms and automatic lighting applications.

2　Facial recognition uses technology and biometrics — typically through AI — to identify human faces. It maps facial features from a photograph or video and then compares the information with a database of known faces to find a match.

quality parameters. Some examples of what can be sensed and monitored include:
- Chemical presence (such as chlorine levels or fluoride levels).
- Oxygen levels (which may impact the growth of algae and bacteria).
- Electrical conductivity (which can indicate the level of ions present in water).
- PH level (a reflection of the relative acidity or alkalinity of the water).
- Turbidity levels (a measurement of the amount of suspended solids in water).

8. Chemical Sensors

Chemical sensors are designed to detect the presence of specific chemical substances. They are often used in production process analysis and environmental pollution monitoring. They are also applied in mineral resources detection, meteorological observation and telemetry, industrial automation, medical distance diagnosis and real-time monitoring, agricultural fresh preservation and fish detection, anti-theft, safety alarm and energy saving.

9. Gas Sensors

Related to chemical sensors, gas sensors are tuned to detect the presence of combustible, toxic, flammable gas in the vicinity of the sensor. Examples of specific gases that can be detected include:
- Bromine
- Carbon monoxide
- Chlorine
- Chlorine dioxide
- Ethylene
- Ethylene oxide
- Formaldehyde
- Hydrazine(s)
- Hydrogen
- Hydrogen bromide
- Hydrogen chloride
- Hydrogen cyanide
- Hydrogen peroxide
- •Hydrogen sulfide
- Nitric oxide
- Nitrogen dioxide
- Ozone
- Peracetic acid
- Propylene oxide
- Sulfur dioxide

10. Smoke Sensors

Smoke sensors or detectors pick up the presence of smoke conditions, which could be an indication of a fire typically, using optical sensors (photoelectric detection) or ionization detection.

11. Infrared (IR) Sensors

Infrared sensor technologies detect infrared radiation that is emitted by objects. Non-contact thermometers make use of these types of sensors as a way of measuring the temperature of an object without having to directly place a probe or sensor on that object. They are used in analyzing the heat signature of electronics and detecting blood flow or blood pressure in patients.

12. Acceleration Sensors

While motion sensors detect movement of an object, acceleration sensors detect the rate of change of velocity of an object. According to the different sensitive components of sensors, common acceleration sensors include capacitive sensors, inductive sensors, strain sensors, piezoresistive sensors, piezoelectric sensors, etc.

13. Gyroscopic Sensors

Gyroscopic sensors measure the rotation of an object and determine the rate of its movement called the angular velocity by using a 3-axis system. These sensors enable the determination of the object's orientation without having to visibly observe it.

14. Humidity Sensors

Humidity sensors are used to measure and monitor the amount of water present in the surrounding air. These sensors are widely used in industries such as semiconductor, biomedical, textiles, food processing, pharmaceuticals, meteorology, microelectronics, agriculture, structural health monitoring, and environment monitoring.

15. Optical Sensors

Optical sensors respond to light that is reflected from an object and generate a corresponding electrical signal. These sensors work by either sensing the interruption of a beam of light or its reflection caused by the presence of the object. Optical sensors include the following types:
- Through-beam sensors (which detect objects by the interruption of a light beam as the object crosses the path between a transmitter and remote receiver).
- Retro-reflective sensors (which combine transmitter and receiver into a single unit and use a separate reflective surface to bounce the light back to the device).
- Diffuse reflection sensors (which operate similarly to retro-reflective sensors except that the object being detected serves as the reflective surface).

New Words

respond   [rɪ'spɒnd]   *v.* 响应；回答

英文	音标	释义
initiative	[ɪˈnɪʃətɪv]	n. 主动性；主动精神；倡议
		adj. 主动的；自发的；创始的
magnitude	[ˈmægnɪtjuːd]	n. 量级
pressure	[ˈpreʃə]	n. 压力；压强
overheat	[ˌəʊvəˈhiːt]	v. 过度加热；（使）变得过热
		n. 过热
ultrasonic	[ˌʌltrəˈsɒnɪk]	adj. 超声的；超声波的，超声速的
		n. 超声波
container	[kənˈteɪnə]	n. 容器
alarm	[əˈlɑːm]	n. 警报；闹铃
capacitive	[kəˈpæsɪtɪv]	adj. 电容性的
illuminate	[ɪˈluːmɪneɪt]	vt. 照亮，照明
ingredient	[ɪnˈɡriːdɪənt]	n. （混合物的）组成部分；原料；要素
parameter	[pəˈræmɪtə]	n. 参数，参量
chlorine	[ˈklɔːriːn]	n. 氯
fluoride	[ˈflɔːraɪd]	n. 氟化物
algae	[ˈældʒiː]	n. 藻类
bacteria	[bækˈtɪərɪə]	n. 细菌（bacterium 的名词复数）
ion	[ˈaɪən]	n. 离子
acidity	[əˈsɪdəti]	n. 酸性
alkalinity	[ˌælkəˈlɪnəti]	n. 碱性
combustible	[kəmˈbʌstəbl]	adj. 易燃的，可燃的
toxic	[ˈtɒksɪk]	adj. 有毒的，中毒的
		n. 毒物，毒剂
flammable	[ˈflæməbl]	adj. 易燃的，可燃的
vicinity	[vəˈsɪnəti]	n. 附近，邻近
bromine	[ˈbrəʊmiːn]	n. 溴
formaldehyde	[fɔːˈmældɪhaɪd]	n. 甲醛；福尔马林
ozone	[ˈəʊzəʊn]	n. 臭氧
detector	[dɪˈtektə]	n. 探测器，检测器
photoelectric	[ˌfəʊtəʊɪˈlektrɪk]	adj. 光电的
ionization	[ˌaɪənaɪˈzeɪʃn]	n. 离子化，电离
radiation	[ˌreɪdɪˈeɪʃn]	n. 辐射，放射物
signature	[ˈsɪɡnətʃə]	n. 识别标志，鲜明特征；签名，署名
rotation	[rəʊˈteɪʃn]	n. 旋转，转动
semiconductor	[ˌsemɪkənˈdʌktə]	n. 半导体
biomedical	[ˌbaɪəʊˈmedɪkl]	adj. 生物医学的
textile	[ˈtekstaɪl]	n. 纺织品，织物；纺织业

pharmaceutical	[ˌfɑːmə'suːtɪkl]	*adj.* 制药的，配药的
meteorology	[ˌmiːtɪə'rɒlədʒi]	*n.* 气象学
microelectronic	[ˌmaɪkrəʊɪˌlek'trɒnɪk]	*adj.* 微电子学的
interruption	[ˌɪntə'rʌpʃn]	*n.* 中断

Phrases

collect data	收集数据
in conjunction with	与……协作
Industrial IoT	工业物联网
temperature sensor	温度传感器
electrical signal	电信号
pressure sensor	压力传感器
atmospheric pressure	大气压力
pressure vessel	压力容器
motion sensor	运动传感器
microwave detection	微波探测
level sensor	液面传感器，液面监测器
closed system	封闭系统
fuel gauge	燃油表
image sensor	图像传感器
license plate	车牌，号码牌
facial recognition	面部识别，人脸识别
proximity sensor	接近传感器
dielectric constant	电容率；介电常数；介质常数
beam of light	光束
electrical conductivity	电导率
turbidity level	浊度
suspended solid	悬浮固体
chemical sensor	化学传感器
chemical substance	化学物质
environmental pollution	环境污染
real-time monitoring	实时监控
fresh preservation	保鲜
gas sensor	气体传感器
carbon monoxide	一氧化碳
chlorine dioxide	二氧化氯
ethylene oxide	环氧乙烷
hydrogen bromide	溴化氢

hydrogen cyanide	氢氰酸
hydrogen peroxide	过氧化氢
hydrogen sulfide	氢化硫
nitric oxide	一氧化氮
peracetic acid	过醋酸，过乙酸
propylene oxide	环氧丙烷
sulfur dioxide	二氧化硫
smoke sensor	烟雾传感器
optical sensor	光学传感器，光敏元件
infrared sensor	红外传感器
blood pressure	血压
acceleration sensor	加速度传感器
capacitive sensor	电容传感器
inductive sensor	感应传感器
strain sensor	应变感传器
piezoresistive sensor	压阻传感器
piezoelectric sensor	压电传感器
gyroscopic sensor	陀螺仪传感器
angular velocity	角速度
humidity sensor	湿度传感器
environment monitoring	环境监测
through-beam sensor	直通波束传感器
retro-reflective sensor	镜面反射型传感器
diffuse reflection sensor	漫反射传感器
reflective surface	反射面

Abbreviations

PIR (Passive Infrared)	被动红外技术
PH (Pondus Hydrogenii)	酸碱度

Exercises

【EX.6】Answer the following questions according to the text.

1. What are sensors?
2. What are sensors designed to do?
3. What do pressure sensors do?
4. What can motion sensors or detectors do?
5. What is a level sensor?
6. What are some examples of what can be sensed and monitored by water quality sensors?

7. What are gas sensors tuned to do?
8. What do infrared sensor technologies do?
9. What are humidity sensors used to do?
10. What types do optical sensors include?

【EX.7】 Translate the following terms or phrases from English into Chinese and vice versa.

1.	detector	1.	
2.	interruption	2.	
3.	real-time monitoring	3.	
4.	parameter	4.	
5.	photoelectric	5.	
6.	semiconductor	6.	
7.	acceleration sensor	7.	
8.	angular velocity	8.	
9.	capacitive sensor	9.	
10.	collect data	10.	
11.	电信号	11.	
12.	面部识别，人脸识别	12.	
13.	感应传感器	13.	
14.	红外传感器	14.	
15.	光学传感器，光敏元件	15.	

【EX.8】 Translate the following sentences into Chinese.
1. An ultrasonic sensor with a battery is put on the back of the glove.
2. We're going to use the money to develop a smaller and lighter ultrasonic sensor.
3. The material is stored in a special radiation proof container.
4. Open system allows users to modify system parameters and initial parameters.
5. Infrared detectors have many uses.
6. Therefore, a various kinds of photoelectric devices have been developed.
7. How do we know that the signature is contemporaneous with the document?
8. Semiconductor devices can perform a variety of control functions in electronic equipment.
9. This paper proposes the method in high-level design of an embedded system interruption controller.
10. The simulation results show validity of presented method for compensating the rotation motion.

Reading Material

IoT Devices

1. What Are IoT Devices, Anyway?

An IoT device is any piece of physical hardware (a "thing," if you will) that's programmed to transmit data over the internet or other networks. IoT technology is often integrated with physical objects, like sensors and appliances. It can also be embedded directly into hardware such as industrial equipment, mobile devices, and other IoT devices.

IoT devices collect data from their environment — things like temperature, heart rate[1], etc.— and exchange that information with other devices and systems in an ecosystem. An IoT-enabled thermostat can detect the room temperature and adjust the heating or air conditioning appropriately, for example. As such, IoT devices create a new dimension[2] of interactions between people and the everyday objects that make up their environment.

2. Three Key Capabilities of IOT Devices

- Sensing — the device has sensors that detect events, changes and conditions in the physical environment like temperature, motion, pressure, location, etc.
- Connectivity — the device can connect to and exchange data over wired or wireless networks. This may be via WiFi, Bluetooth, cellular, satellite or other communication protocols.
- Data exchange — the device can send sensor data over networks and in some cases also receive data and commands to actuators that can control mechanisms and systems.

3. The Core Components of an IoT Device

- Sensors and actuators — sensors and actuators detect events and conditions and enable responses. Sensors convert physical properties into electrical signals[3]. Actuators convert electrical signals into physical actions.
- Processors — processors execute code that processes sensor data and controls actuators. They range from basic microcontrollers to advanced microprocessors and systems-on-chip[4].
- Communication hardware — communication hardware is connected to wired and wireless networks. It include chips, antennas, ports, etc.
- Software — the device has firmware and applications to manage device operation, data exchange protocols, process data and enable communication.

Operating system — some devices have compact real-time operating systems suited for IoT devices.

---

1 heart rate：心率
2 dimension [daɪˈmenʃn] n. 维度
3 electrical signal：电子信号
4 systems-on-chip：片载系统，片上系统

- Security features — the device has hardware and software to secure device communications and data transmissions.
- Power supply — the device has battery, power harvesting or AC power connection.

IoT devices take input from the physical world through sensors, process it and transmit it over networks. The connectivity allows them to exchange data with applications, services, other devices and users to enable useful functions.

4. How Do IoT Devices Work?

The basic function of any IoT device, regardless of the context or application, can be boiled down to[1] one fundamental goal: to collect data and share it across a network. Here's how that works.

IoT devices are connected to a network. In this way, they can communicate and interact with each other without actually establishing a physical connection. The network they communicate with may include cellular, satellite, WiFi, Bluetooth, low power wide area networks (LPWAN)[2], or an ethernet cable. The most popular networks used, however, are WiFi and Bluetooth. Some devices are accessible directly over the public internet, but the majority are integrated over a local private network for security purposes.

Data is collected by the devices and sent over the network to an IoT gateway or other edge device for centralized data storage. Depending on the application, that might look like reading air quality from a pollution sensor or recording a live video for a smart security system, for example.

Next, the data is analyzed locally on the centralized device, or sent to the cloud for processing. At this point, the centralized device can synchronize the behavior of the other edge devices in the network, or orchestrate[3] complex actions like making decisions based on what's happening to the system as a whole. An example of this is turning on all of the lights outside when motion is detected from a single sensor, or calling emergency services when an edge device sends a specific signal.

Finally, the data is made available to the end user in a way that's easily understood (if the interface is designed well). That could look like receiving live video feeds directly to your phone from a remote monitoring camera, or getting a text alert when temperatures get too high in a certain area.

5. The Benefits of IoT Devices

IoT devices are designed to make life easier and to make processes more intelligent. The benefits of integrating IoT devices into your business are many, including:
- Automation and control over otherwise manual data collection and processes.

1　be boiled down to：归结为
2　low power wide area networks：低功耗广域网
3　orchestrate ['ɔ:kɪstreɪt] vt. 协调

- Real-time monitoring capabilities at scale for any number of devices.
- Improved productivity by reducing manual tasks.
- Enabling predictive maintenance for machines and parts to reduce downtime.
- Leveraging data and machine learning to increase operational efficiency and management.
- Scheduled maintenance[1] can ensure compliance with required regulations that are traditionally prone to human error.
- Enhanced customer experience through:customization and personalization of home and personal technology, various user interfaces (email, text, App notification), understanding customer behavior, improving customer service through alerting to issues on a device, notifying customers of needed maintenance, identifying end of warranty dates, etc.

6. What Does IoT Device Management Mean, and Why Does It Matter?

Implementing IoT technology can quickly become a highly complex and resource intensive[2] operation, and will only grow in complexity and scale as more is invested into the technology.

Just as smartphones and laptops need regular updates for security, features and bugs, your IoT devices will require updates as well. To keep the system running at peak capacity, each individual IoT device requires regular management throughout its lifecycle to maintain device health, connectivity, and security.

IoT device management refers to the processes, approach, and tooling that help remotely handle the complex infrastructure of interconnected IoT devices. Successful IoT device management starts by planning for the entire life cycle of devices and applications. After all, these solutions are intended to become an integral part of an enterprise's technical ecosystem over several years.

This means that it's critical to develop and maintain a robust IoT device management strategy, or to partner with a trusted IoT technology provider to oversee this strategy for your business.

7. Device Management Best Practices

A complex IoT device infrastructure requires a supportive[3] power grid, reliable communication network access, data processing, and storage capabilities. Devices need to be registered, configured, integrated, maintained, and monitored. Additionally, devices need to uphold industry security standards to defend against breaches and protect private consumer data.

The elements of this device management system and lifecycle include:

- Initial provisioning[4] of device software and configuration so as not to require user input.
- Registering devices in the system before they are actually connected.

1　scheduled maintenance：定期维修，例行维修
2　resource intensive：资源密集型
3　supportive [sə'pɔːtɪv] adj. 支持的
4　provision [prə'vɪʒn] n. 配置；预备，准备

- Dynamic partitioning[1] for making changes to software and configuration without disrupting[2] functionality.
- Remote monitoring and diagnostics[3] in a centralized dashboard where all device data and behavior can be observed.
- Bulk[4] device management through dynamic hierarchies and logical groupings to increase deployment and maintenance efficiency.
- Remote configuration of devices already in use by the system.
- Software and firmware updates via continuous integration and continuous deployment (CI/CD)[5] pipelines.
- Integration & extensibility for access to robust APIs and interfacing with platforms through software development kits (SDKs)[6].
- Decommissioning[7] old IoT devices that are no longer in use.

8. IoT Device Security

IoT devices are no stranger to security breaches, which is why the importance of IoT security can't be overstated. Malicious actors have used various vulnerable[8] IoT devices — from from internet-connected gas pumps to medical devices — to infiltrate[9] unwitting victims, launch launch botnets, and cause millions of dollars in damages.

IoT device security presents a unique challenge due to the sheer number of potential vulnerabilities. Your teams will need to secure:

- The devices themselves.
- The devices' firmware.
- Any applications that interact with the devices.
- The cloud infrastructure that hosts or supports the product.

参考译文

Text A 物联网协议

1. MQTT

MQTT 是物联网安全领域广泛采用的安全协议。MQTT 代表消息队列遥测传输,是一

---

1 dynamic partitioning:动态分区
2 disrupt [dɪsˈrʌpt] vt. 破坏,使中断
3 diagnostic [ˌdaɪəɡˈnɒstɪk] adj. 诊断的
4 bulk [bʌlk] adj. 大批的,大量的
5 continuous integration and continuous deployment (CI/CD):持续集成和持续部署
6 software development kits (SDKs):软件开发工具包
7 decommission [ˌdiːkəˈmɪʃn] vt. 使退役
8 vulnerable [ˈvʌlnərəbl] adj. 易受攻击的
9 infiltrate [ˈɪnfɪltreɪt] v.(使)潜入,(使)渗透,(使)渗入

种客户端-服务器通信消息传输协议。它通过 TCP/IP 或其他提供可靠、无损和双向连接的协议运行。

MQTT 的特点如下：

MQTT 是一种轻量级且简单的协议，有助于快速、高效的数据传输。它专门为低带宽、高延迟或不可靠的受限设备和网络而设计。

该协议使用最少的数据包，从而减少了网络使用量，而其最佳功耗有助于延长连接设备的电池寿命，使其成为移动电话和可穿戴设备的理想选择。

MQTT 基于消息传递技术，可确保快速、可靠的通信。因此，它非常适合在物联网应用中使用。

它用在哪里：

MQTT 的安全性分为不同的层，即网络层、传输层和应用程序层，每层都用于阻止特定形式的攻击。鉴于 MQTT 本质上是一种轻量级协议，因此它包含的安全机制的数量有限。为了增强安全性，实施 MQTT 时经常会利用其他安全标准，例如用于传输加密的 SSL/TLS、以确保网络物理安全的网络级别的 VPN，以及用户名/密码。此外，客户端标识符随数据包一起传输，以便在应用程序级别对设备进行身份验证。

2. CoAP

CoAP（约束应用程序协议）是一种网络传输协议，专门设计用于满足受约束设备（如微控制器）及其运行的低功耗或有损网络的要求。它被广泛认为是保护物联网应用程序最流行的协议之一。

CoAP 的特点如下：

与 HTTP 一样，CoAP 建立在 REST 架构上。

客户端使用 GET、PUT、POST 和 DELETE 等方法通过 URL 访问服务器提供的资源。

CoAP 专门设计用于在微控制器上运行。它是物联网的理想协议，需要数百万个低成本节点。

CoAP 资源高效，需要的设备和网络上的资源最少。它在 IP 上使用 UDP，而不是复杂的传输堆栈。

它用在哪里：

CoAP 利用用户数据报模型（UDP）进行信息传输，因此依赖 UDP 的安全性来保护信息。CoAP 采用基于 UDP 的数据报传输层安全性（DTLS）来增强安全性。

CoAP 的界面简单且用户友好，可与 HTTP 无缝集成以进行网络集成。它还提供多播支持和低开销等功能，从而有助于物联网的安全。

3. DTLS

DTLS 协议是专门为物联网开发的安全措施，用于保护依赖数据报的应用程序之间的数据通信。DTLS 建立在传输层安全（TLS）协议的基础上，它提供同等级别的安全性。

DTLS 的主要目标是通过对 TLS 进行细微修改来应对数据丢失和重新排序等挑战。DTLS 协议保留了底层传输层的语义，从而避免了由相关流协议引起的任何延迟。然而，应用程序必须处理数据报丢失、数据包重新排序及数据超出数据报网络数据包大小等问题。

DTLS 的特点如下：

DTLS 采用重传计时器来解决数据包丢失的问题。

如果计时器在客户端收到服务器的确认消息之前已经到期，那么客户端将重新传输数据。为了缓解重新排序的问题，每个消息都被分配了一个唯一的序列号，从而可以确定后续收到的消息是否按顺序排列。如果不按顺序排列，则将其放入队列中，并在获得序列号时进行处理。

它用在哪里：

DTLS 常用于各种应用，包括实时视频源、视频流、游戏、VoIP 和即时消息。该协议特别适合低延迟比数据丢失更重要的情况下。

4. 6LoWPAN

6LoWPAN 协议代表低功耗无线个人区域网上的 IPv6，专为无线传感器网络和物联网系统等低功耗网络而设计。

6LoWPAN 的特点如下：

6LoWPAN 是一种用于在不同网络上以 IPv6 形式传输数据包的协议。它提供端到端 IPv6 连接，从而可以直接访问包括互联网在内的各种网络。此外，6LoWPAN 用于保护最终用户和传感器网络之间的通信。

为了确保物联网的安全性，6LoWPAN 采用 IEEE 802.15.4 中定义的 AES-128 链路层安全性。利用链路验证和加密来提供安全性，并为通过 TCP 运行的传输层安全机制提供进一步的安全性。

它用在哪里：

6LoWPAN 是智能家居自动化、工业监控、智能电网和通用自动化等各个领域的关键技术。

5. ZigBee

ZigBee 被广泛认为是一种尖端协议，可为物联网设备和应用程序提供强大的安全性。该技术有助于 10~100m 距离内机器对机器的无缝通信，使其成为无线电系统等低功耗嵌入式设备的理想选择。此外，ZigBee 是一种开源无线技术，它既经济又高效。

ZigBee 的特点如下：

ZigBee 提供跨所有层的标准化，促进了不同制造商的产品之间的兼容性。其网状架构有利于与附近设备的连接，从而扩展网络并增强其灵活性。

ZigBee "绿色能源"的实施降低了能源消耗和成本。此外，ZigBee 支持大量设备（大约 6550 个），有助于网络的可扩展性。

它用在哪里：

ZigBee 主要应用于家庭自动化、医疗数据采集、工业控制系统、抄表系统、灯光控制系统、商业、全球政府市场、家庭网络等。

6. AMQP

AMPQ 是一种高效、可移植、多通道消息传递的协议，优先考虑安全性。该协议通过 SASL 或 TLS 提供身份验证和加密，这些协议依赖 TCP 等传输协议。

AMQP 的特点如下：
AMQP 协议的开发目的是促进各种应用程序和系统之间的通信，无论其内部架构如何。这促成了工业规模的商业通信标准化。

它用在哪里：
该协议用于客户端/服务器通信及物联网设备的管理。AMPQ 以其高效、轻便、多通道功能和安全功能而自豪。

7. DDS

DDS 是一种发布-订阅协议，与 MQTT 不同，它不需要连接服务器。相反，DDS 采用无代理架构，从而形成不依赖任何中间系统的高速、高性能协议。DDS 由对象管理组织（OMG）开发，专为设备到设备通信而设计。

DDS 的特点如下：
DDS 技术有助于创建具有模块化和松耦合的开放式体系结构的系统。

它通过在子系统和组件之间启用明确定义的接口来实现这一点，从而消除封闭和专有的架构。

这种方法降低了集成、维护和升级的成本，同时在中间件和子系统级别上促进竞争和易重用性。

此外，DDS 标准化了消息语义，增强了系统的健壮性并降低了整体开发和集成的成本。

它用在哪里：
DDS 满足了航空航天、国防、空中交通管制、自动驾驶汽车、医疗设备、机器人、模拟和测试、智能电网管理、交通系统和其他相关领域的各种应用的实时数据交换需求。

Unit 4

Text A

Network Architecture

Network architecture is the design of a communications network. It is a framework for the specification of a network's physical components and their functional organization and configuration, its operational principles and procedures, as well as data format sued in its operation.

In telecommunication, the specification of network architecture may also include a detailed description of products and services delivered via a communications network, as well as detailed rate and billing structures under which services are.

1. OSI Network Model

The open systems interconnection model[1] (OSI model) is a product of the open systems interconnection effort at the International Organization for Standardization. It is a way of sub-dividing a communications system into smaller parts called layers. A layer is a collection of similar functions that provide services to the layer above it and receives services from the layer below it. On each layer, an instance provides services to the instances at the layer above and requests service from the layer below.

1.1 Physical Layer

The physical layer defines the electrical and physical specifications for devices. In particular, it defines the relationship between a device and a transmission medium, such as a copper or optical cable. This includes the layout of pins, voltages, cable specifications, hubs, repeaters,

---

1 The open systems interconnection (OSI) model (ISO/IEC 7498-1) is a product of the open systems interconnection effort at the International Organization for Standardization. It is a prescription（[prɪˈskrɪpʃn] n. 指示，规定）of characterizing and standardizing the functions of a communications system in terms of abstraction layers.

network adapters, host bus adapters (HBA used in storage area networks) and more. Its main task is the transmission of a stream of bits over a communication channel.

1.2 Data Link Layer

The data link layer provides the functional and procedural means to transfer data between network entities and to detect and possibly correct errors that may occur in the physical layer. Originally, this layer was intended for point-to-point and point-to-multipoint media, characteristic of wide area media in the telephone system. Local area network architecture, which included broadcast-capable multi-access media, was developed independently of the ISO work in IEEE Project 802. In modern practice, only error detection, not flow control using sliding window, is present in data link protocols such as point-to-point protocol (PPP)[1], and, on local area networks, the IEEE 802.2 LLC[2] layer is not used for most protocols on the Ethernet, and on other local area networks, its flow control and acknowledgment mechanisms are rarely used. Sliding window flow control and acknowledgment is used at the transport layer by protocols such as TCP, but is still used in niches where X.25[3] offers performance advantages. Simply, its main job is to create and recognize the frame boundary. This can be done by attaching special bit patterns to the beginning and the end of the frame. The input data is broken up into frames.

1.3 Network Layer

The network layer provides the functional and procedural means of transferring variable length data sequences from a source host on one network to a destination host on a different network, while maintaining the quality of service requested by the transport layer (in contrast to the data link layer which connects hosts within the same network). The network layer performs network routing functions, and might also perform fragmentation and reassembly, and report delivery errors. Routers operate at this layer—sending data throughout the extended network and making the Internet possible. This is a logical addressing scheme—values are chosen by the network engineer. The addressing scheme is not hierarchical. It controls the operation of the subnet and determines the routing strategies between IMPs[4] and insures that all the packs are correctly received at the destination in the proper order.

1.4 Transport Layer

The transport layer provides transparent transfer of data between end users, providing

---

1 In networking, the point-to-point protocol (PPP) is a data linkprotocol commonly used in establishing a direct connection between two networking nodes. It can provide connection authentication, transmission encryption, and <u>compression</u>（[kəmˈpreʃən] *n.* 压缩）.

2 In the seven-layer OSI model of computer networking, the logical link control (LLC) data communication protocol layer is the upper sublayer of the data link layer, which is itself layer 2.

3 X.25 is an ITU-T standard protocol suite for packet switchedwide area network (WAN) communication. An X.25 WAN consists of packet-switching exchange (PSE) nodes as the networking hardware, and <u>leased lines</u>（租用线，专用线）, plain old telephone service connections or ISDN connections as physical links.

4 The interface message processor (IMP) was the packet-switching node used to interconnect <u>participant</u>（[pɑːˈtɪsɪpənt] *adj.* 参与的）networks to the ARPANET from the late 1960s to 1989. It was the first generation of gateways, which are known today as routers.

reliable data transfer services to the upper layers. The transport layer controls the reliability of a given link through flow control, segmentation/desegmentation, and error control. Some protocols are state and connection oriented. This means that the transport layer can keep track of the segments and retransmit those that fail. The transport layer also provides the acknowledgement of the successful data transmission and sends the next data if no errors occurred. Some transport layer protocols, for example TCP, but not UDP, support virtual circuits[1] and provide connection oriented communication over an underlying packet oriented datagram[2] network. The datagram transportation delivers the packets randomly and broadcasts it to multiple nodes.

Notes: the transport layer multiplexes several streams on to 1 physical channel. The transport header tells which message belongs to which connection.

1.5 The Session Layer

This layer provides a user interface to the network where the user negotiates to establish a connection. The user must provide the remote address to be contacted. The operation of setting up a session between two processes is called "binding". In some protocols it is merged with the transport layer. Its main work is to transfer data from the other application to this application so this application is mainly used for transferred layer.

1.6 Presentation Layer

The presentation layer establishes context between application layer entities, in which the higher-layer entities may use different syntax and semantics if the presentation service provides a mapping between them. If a mapping is available, presentation service data units are encapsulated into session protocol data units, and passed down the stack. This layer provides independence from data representation (e.g., encryption) by translating between application and network formats. The presentation layer transforms data into the form that the application accepts. This layer formats and encrypts data to be sent across a network. It is sometimes called the syntax layer. The original presentation structure uses the basic encoding rules of abstract syntax notation one (ASN.1), with capabilities such as converting an EBCDIC-coded text file to an ASCII-coded file, or serialization of objects and other data structures from and to XML.

1.7 Application Layer

The application layer is the OSI layer closest to the end user, which means that both the OSI application layer and the user interact directly with the software application. This layer interacts with software applications that implement a communicating component. Such application programs fall outside the scope of the OSI model. Application layer functions typically include

---

1　In telecommunications and computer networks, a virtual circuit (VC), synonymous with virtual connection and virtual channel, is a connection oriented communication service that is delivered by means of packet mode communication. After a connection or virtual circuit is established between two nodes or application processes, a <u>bit stream</u>（位流）or byte stream may be delivered between the nodes; a virtual circuit protocol allows higher level protocols to avoid dealing with the division of data into segments, packets, or frames.

2　A datagram is a basic transfer unit associated with a packet-switched network in which the delivery, arrival time, and order of arrival are not guaranteed by the network service.

identifying communication partners, determining resource availability, and synchronizing communication. When identifying communication partners, the application layer determines the identity and availability of communication partners for an application with data to transmit.

2. Distributed Computing

In distributed computing, the term network architecture often describes the structure and classification of distributed application architecture, as the participating nodes in a distributed application are often referred to as a network. For example, the applications architecture of the public switched telephone network (PSTN)[1] has been termed the advanced intelligent network[2]. There are a number of specific classifications but all lie on a continuum between the dumb network[3] (e.g., Internet) and the intelligent computer network (e.g., the telephone network). Other networks contain various elements of these two classical types to make them suitable for various types of applications. Recently the context aware network[4], which is a synthesis of two, has gained much interest with its ability to combine the best elements of both.

A popular example of such usage of the term in distributed applications as well as PVCs (permanent virtual circuits) is the organization of nodes in peer-to-peer (P2P) services and networks. P2P networks usually implement overlay networks running over an underlying physical or logical network. These overlay network may implement certain organizational structures of the nodes according to several distinct models, the network architecture of the system.

New Words

architecture	['ɑːkɪtektʃə]	n. 体系，机构
framework	['freɪmwɜːk]	n. 构架，框架，结构
specification	[ˌspesɪfɪ'keɪʃn]	n. 规范，详述，规格，说明书
functional	['fʌŋkʃənl]	adj. 功能的

1　The public switched telephone network (PSTN) is the network of the world's public circuit-switched telephone networks. It consists of telephone lines, fiber optic cables, <u>microwave</u> (['maɪkrəweɪv] n. 微波) transmission links, cellular networks, communications satellites, and undersea telephone cables, all inter-connected by switching centers, thus allowing any telephone in the world to communicate with any other.

2　The intelligent network (IN) is the standard network architecture specified in the ITU-T Q.1200 series recommendations. It is intended for fixed as well as mobile telecom networks. It allows operators to differentiate themselves by providing <u>value-added services</u>（增值服务）in addition to the standard telecom services such as PSTN, ISDN and <u>GSM</u>（Global System of Mobile communication，全球移动通信系统）services on mobile phones.

3　A dumb network is marked by using intelligent devices (i.e. PCs) at the periphery that make use of a network that does not interfere or manage with an application's operation / communication. The dumb network concept is the natural outcome of theend to end principle. The Internet was originally designed to operate as a dumb network.

4　A context aware network is a network that tries to overcome the limitations of the dumb and intelligent network models and to create a <u>synthesis</u> (['sɪnθəsɪs] n. 综合) which combines the best of both network models. It is designed to allow for customization and application creation while at the same time ensuring that application operation is compatible not just with the preferences of the individual user but with the expressed preferences of the enterprise or other <u>collectivity</u> ([ˌkɒlek'tɪvɪtɪ] n. 全体，总体) which owns the network. The Semantic Web is an example of a context aware network.

configuration	[kənˌfɪɡəˈreɪʃn]	n. 构造，结构，配置
principle	[ˈprɪnsəpl]	n. 法则，原则，原理
procedure	[prəˈsiːdʒə]	n. 进程，程序
layer	[ˈleɪə]	n. 层
collection	[kəˈlekʃn]	n. 收集；聚集
instance	[ˈɪnstəns]	n. 实例，要求
		vt. 举……为例
request	[rɪˈkwest]	vt. & n. 请求，要求
repeater	[rɪˈpiːtə]	n. 转发器，中继器
storage	[ˈstɔːrɪdʒ]	n. 存储
entity	[ˈentətɪ]	n. 实体
multipoint	[ˌmʌltɪˈpɔɪnt]	adj. 多点（式）的，多位置的
characteristic	[ˌkærəktəˈrɪstɪk]	adj. 特有的，表示特性的，典型的
		n. 特性，特征
broadcast	[ˈbrɔːdkɑːst]	n. & v. 广播
Ethernet	[ˈiːθənet]	n. 以太网
acknowledgment	[əkˈnɒlɪdʒmənt]	n. 确认，承认
niche	[niːʃ]	n. 合适的位置，小生境
performance	[pəˈfɔːməns]	n. 履行，执行，性能
sequence	[ˈsiːkwəns]	n. 次序，顺序，序列
fragmentation	[ˌfræɡmenˈteɪʃn]	n. 分段
reassembly	[ˌriːəˈsemblɪ]	n. 重新装配
router	[ˈruːtə]	n. 路由器
hierarchical	[ˌhaɪəˈrɑːkɪkl]	adj. 分等级的
subnet	[ˈsʌbnet]	n. 子网络，分支网络
transparent	[trænsˈpærənt]	adj. 透明的，显然的，明晰的
segmentation	[ˌseɡmenˈteɪʃn]	n. 分割
connection	[kəˈnekʃn]	n. 连接，接线，线路
retransmit	[rɪtrænzˈmɪt]	v. 转播，转发
datagram	[ˈdeɪtəɡræm]	n. 数据报
multiplex	[ˈmʌltɪpleks]	v. 多路传输，多路复用；多重发信
		adj. 多元的
establish	[ɪˈstæblɪʃ]	vt. 建立，设立
session	[ˈseʃn]	n. 会话
binding	[ˈbaɪndɪŋ]	n. 绑定
semantics	[sɪˈmæntɪks]	n. 语义
mapping	[ˈmæpɪŋ]	n. 映射
map	[ˈmæp]	vt. 映射

encapsulate	[ɪnˈkæpsjuleɪt]	v. 封装
stack	[stæk]	n. 堆栈
		v. 堆叠
format	[ˈfɔːmæt]	n. 形式，格式
		vt. 安排……的格式
encrypt	[ɪnˈkrɪpt]	v. 加密，将……译成密码
convert	[kənˈvɜːt]	vt. 使转变，转换……
serialization	[ˌsɪərɪəlaɪˈzeɪʃn]	n. 序列化
synchronize	[ˈsɪŋkrənaɪz]	v. 同步
identity	[aɪˈdentəti]	n. 身份，一致
continuum	[kənˈtɪnjuəm]	n. 连续统一体，闭联集
synthesis	[ˈsɪnθəsɪs]	n. 综合，合成

Phrases

communication network	通信网络
open systems interconnection model (OSI model)	开放系统互联参考模型
International Organization for Standardization	国际标准化组织
physical layer	物理层
transmission medium	传输介质，传送介质
optical cable	光缆
stream of bits	比特流
data link layer	数据链路层
intend for…	打算供……使用
local area network	局域网
independent of…	不依赖……，独立于……
point-to-point protocol (PPP)	点对点协议
flow control	流控制
sliding window	滑动窗口
transport layer	传输层
bit pattern	位组合格式，位的形式
network layer	网络层
in contrast to …	和……形成对比，和……形成对照
logical addressing scheme	逻辑寻址方案
virtual circuit	虚拟线路，虚拟电路
session layer	会话层
presentation layer	表示层
abstract syntax notation one	抽象语法表示法1，抽象语法符号1
application layer	应用层

fall outside	超出……，超越……
distributed computing	分布式计算
advanced intelligent network	高级智能网
suitable for…	适合……的
context aware network	情景感知网络
overlay network	覆盖网络，重叠网络，叠加网络

Abbreviations

HBA (Host Bus Adapter)	主机总线适配器
LLC (Logical Link Control)	逻辑链路控制
IMP (Interface Message Processor)	接口信息处理器
UDP (User Datagram Protocol)	用户数据报协议
EBCDIC (Extended Binary Coded Decimal Interchange Code)	扩充的二进制编码的十进制交换码
ASCII (American Standard Code for Information Interchange)	美国信息交换标准码
XML (eXtensible Markup Language)	可扩展标记语言
PVC (Permanent Virtual Circuit)	永久虚电路

Analysis of Difficult Sentences

[1] A layer is a collection of similar functions that provide services to the layer above it and receives services from the layer below it.

本句中，that provide services to the layer above it and receives services from the layer below it 是一个定语从句，修饰和限定 similar functions。provide sth. to sb.的意思是"为某人提供某物"。

[2] Local area network architecture, which included broadcast-capable multi-access media, was developed independently of the ISO work in IEEE Project 802.

本句中，which included broadcast-capable multi-access media 是一个非限定性定语从句，对 Local area network architecture 进行补充说明。independently of 的意思是"独立于；与……无关"。

[3] The network layer provides the functional and procedural means of transferring variable length data sequences from a source host on one network to a destination host on a different network, while maintaining the quality of service requested by the transport layer (in contrast to the data link layer which connects hosts within the same network).

本句中，while maintaining the quality of service requested by the transport layer (in contrast to the data link layer which connects hosts within the same network)是时间状语，while 表示"在……的同时"。

英语中，当 while 引导时间状语从句时，它可以表示"当某事发生时，另一件事也在同一时刻发生"。这种用法强调的是两件事情同时发生。例如：

While waiting for the boss, he was reading the product description.

他一边等老板,一边读着产品说明书。

[4] The presentation layer establishes context between application layer entities, in which the higher-layer entities may use different syntax and semantics if the presentation service provides a mapping between them.

本句中,in which the higher-layer entities may use different syntax and semantics if the presentation service provides a mapping between them 是一个介词前置的非限定性定语从句,对 application layer entities 进行补充说明。在该从句中,if the presentation service provides a mapping between them 是一个条件状语从句,修饰谓语 may use。

[5] The application layer is the OSI layer closest to the end user, which means that both the OSI application layer and the user interact directly with the software application.

本句中,closest to the end user 是形容词短语,作后置定语,修饰和限定 the OSI layer,它可以扩展为一个定语从句:which is closest to the end user。which means that both the OSI application layer and the user interact directly with the software application 是一个非限定性定语从句,对它前面的整个句子进行补充说明。

Exercises

【EX.1】 Answer the following questions according to the text.

1. What is the open systems interconnection model (OSI model)?
2. What is a layer?
3. What does the physical layer define?
4. What does the data link layer provide?
5. What does the networ layer provide?
6. What does the transport layer control?
7. What is binding?
8. What does the presentation layer establish?
9. What do application layer functions typically include?
10. What does the term network architecture often describe in distinct usage in distributed computing?

【EX.2】 Translate the following terms or phrases from English into Chinese and vice versa.

1.	data link layer	1.	
2.	point-to-point protocol (PPP)	2.	
3.	transport layer	3.	
4.	logical addressing scheme	4.	
5.	local area network	5.	
6.	session layer	6.	
7.	distributed computing	7.	

8.	virtual circuit	8.	
9.	presentation layer	9.	
10.	data link layer	10.	
11.	*n.* 构架，框架，结构	11.	
12.	*adj.* 功能的	12.	
13.	*n.* 转发器，中继器	13.	
14.	*n.* 进程，程序	14.	
15.	*n.* 路由器	15.	

【EX.3】 **Translate the following sentences into Chinese.**

1. It is certified that the software architecture is universal and expandable.
2. In an object-oriented environment, a framework consists of abstract and concrete classes.
3. The software does something that the product specification says it shouldn't do.
4. An error has occurred during configuration of home networking on this computer.
5. A hub is a device that connects several nodes of a local area network.
6. The signal transmitted between the end facilities is relayed by the repeater.
7. You cannot disable the selected network adapter because it is used for network connectivity.
8. Only Ethernet facilitates the combination of components from different manufacturers.
9. Which command will configure a default route on a router?
10. Classifying the datagram on LANs can provide some evidences of NID (Network Intrusion Detect).

【EX.4】 **Complete the following passage with appropriate words in the box.**

same	workstation	collision	decreased	captured
delay	token-passing	system	logical	node

The Ethernet protocol is by far the most widely used. Ethernet uses an access method called CSMA/CD (Carrier Sense Multiple Access/Collision Detection). This is a ___1___ where each computer listens to the cable before sending anything through the network. If the network is clear, the computer will transmit. If some other ___2___ is already transmitting on the cable, the computer will wait and try again when the line is clear. Sometimes, two computers attempt to transmit at the ___3___ instant. When this happens a ___4___ occurs. Each computer then backs off and waits a random amount of time before attempting to retransmit. With this access method, it is normal to have collisions. However, the ___5___ caused by collisions and retransmitting is very small and does not normally effect the speed of transmission on the network.

The Ethernet protocol allows for linear bus, star, or tree topologies. Data can be transmitted over wireless access points, twisted pair, coaxial, or fiber optic cable at a speed of 10 Mbps up to 1000 Mbps.

The token ring protocol was developed by IBM in the mid-1980s. The access method used involves ___6___. In token ring, the computers are connected so that the signal travels around the network from one computer to another in a ___7___ ring. A single electronic token moves around the ring from one computer to the next. If a computer does not have information to transmit, it simply passes the token on to the next ___8___. If a computer wishes to transmit and receives an empty token, it attaches data to the token. The token then proceeds around the ring until it comes to the computer for which the data is meant. At this point, the data is ___9___ by the receiving computer. The token ring protocol requires a star-wired ring using twisted pair or fibre optic cable. It can operate at transmission speeds of 4 Mbps or 16 Mbps. Due to the increasing popularity of Ethernet, the use of token ring in school environments has ___10___.

【EX.5】 **Translate the following passage into Chinese.**

Network

A network is a group of two or more computer systems linked together. There are many types of computer networks, including:

- Local area networks (LANs): the computers are geographically close together.
- Wide area networks (WANs): the computers are farther apart and are connected by telephone lines or radio waves.
- Campus area networks (CANs): the computers are within a limited geographic area, such as a campus or military base.
- Metropolitan area networks (MANs): a data network designed for a town or city.
- Home area networks (HANs): a network contained within a user's home that connects a person's digital devices.

In addition to these types, the following characteristics are also used to categorize different types of networks:

- Topology: the geometric arrangement of a computer system. Common topologies include a bus, star, and ring.
- Protocol : the protocol defines a common set of rules and signals that computers on the network use to communicate. One of the most popular protocols for LANs is called Ethernet. Another popular LAN protocol for PCs is the IBM token-ring network.
- Architecture: networks can be broadly classified as using either a peer-to-peer or client/server architecture.

Computers on a network are sometimes called nodes. Computers and devices that allocate resources for a network are called servers.

Text B

Network Topology

Network topology is the arrangement of the various elements (links, nodes, etc.) of a computer or biological network. Essentially, it is the topological structure of a network, and may be depicted physically or logically. Physical topology refers to the placement of the network's various components, including device location and cable installation, while logical topology shows how data flows within a network, regardless of its physical design. Distances between nodes, physical interconnections, transmission rates, and/or signal types may differ between two networks, yet their topologies may be identical.

A good example is a local area network (LAN)[1]: any given node in the LAN has one or more physical links to other devices in the network; graphically mapping these links results in a geometric shape that can be used to describe the physical topology of the network. Conversely, mapping the data flow between the components determines the logical topology of the network.

There are two basic categories of network topologies:
- Physical topologies.
- Logical topologies.

The shape of the cabling layout used to link devices is called the physical topology of the network. This refers to the layout of cabling, the locations of nodes, and the interconnections between the nodes and the cabling. The physical topology of a network is determined by the capabilities of the network access devices and media, the level of control or fault tolerance desired, and the cost associated with cabling or telecommunications circuits.

The logical topology, in contrast, is the way that the signals act on the network media, or the way that the data passes through the network from one device to the next without regard to the physical interconnection of the devices. A network's logical topology is not necessarily the same as its physical topology. For example, the original twisted pair Ethernet[2] using repeater hubs was a logical bus topology with a physical star topology layout. Token ring is a logical ring topology, but is wired a physical star from the media access unit[3].

The logical classification of network topologies generally follows the same classifications as those in the physical classifications of network topologies but describes the path that the data takes between nodes being used. The logical topologies are generally determined by network protocols.

---

1 A local area network (LAN) is a computer network that interconnects computers in a limited area such as a home, school, computer laboratory, or office building using network media.

2 The twisted pair Ethernet technologies use twisted-pair cables for the physical layer of an Ethernet computer network. Other Ethernet cable standards employ coaxial cable or optical fiber.

3 A media access unit (MAU, also called multistation access unit, MSAU) is a device to attach multiple network stations in a star topology in a token ring network, internally wired to connect the stations into a logical ring.

Logical topologies are often closely associated with media access control[1] methods and protocols. Logical topologies are able to be dynamically reconfigured by special types of equipment such as routers and switches.

1. Point-to-point

The simplest topology is a permanent link between two endpoints. Switched point-to-point topologies are the basic model of conventional telephony. The value of a permanent point-to-point network is unimpeded communications between the two endpoints. The value of an on-demand point-to-point connection is proportional to the number of potential pairs of subscribers, and has been expressed as Metcalfe's Law[2].

Permanent (dedicated): easiest to understand, of the variations of point-to-point topology, is a point-to-point communications channel that appears, to the user, to be permanently associated with the two endpoints. A children's tin can telephone is one example of a physical dedicated channel.

Within many switched telecommunications systems, it is possible to establish a permanent circuit. One example might be a telephone in the lobby of a public building, which is programmed to ring only the number of a telephone dispatcher. "Nailing down" a switched connection saves the cost of running a physical circuit between the two points. The resources in such a connection can be released when no longer needed, for example, a television circuit from a parade route back to the studio.

Switched: using circuit switching[3] or packet switching[4] technologies, a point-to-point circuit can be set up dynamically, and dropped when no longer needed. This is the basic mode of conventional telephony.

2. Bus

In local area networks where bus topology is used, each node is connected to a single cable (see Figure 4-1). Each computer or server is connected to the single bus cable. A signal from the source travels in both directions to all machines connected on the bus cable until it finds the intended recipient. If the machine address does not match the intended address for the data, the

---

1 In the seven-layer OSI model of computer networking, media access control (MAC) data communication protocol is a <u>sublayer</u> (['sʌb'leɪə] *n.*下层) of the data link layer, which itself is layer 2. The MAC sublayer provides addressing and channel access control mechanisms that make it possible for several terminals or network nodes to communicate within a multiple access network that incorporates a shared medium, e.g. Ethernet.

2 Metcalfe's law states that the value of a telecommunications network is <u>proportional</u> ([prə'pɔːʃənl] *adj.*成比例的) to the square of the number of connected users of the system (n^2).

3 Circuit switching is a methodology of implementing a telecommunications network in which two network nodes establish a dedicated communications channel (circuit) through the network before the nodes may communicate. The circuit <u>guarantees</u> ([ˌgærən'tiːz] *vt.* 保证) the full bandwidth of the channel and remains connected for the duration of the communication session. The circuit functions as if the nodes were physically connected as with an electrical circuit.

4 Packet switching is a digital networking communications method that groups all transmitted data – <u>regardless of</u>(不管，不顾) content, type, or structure – into suitably sized blocks, called packets.

machine ignores the data. Alternatively, if the data matches the machine address, the data is accepted. Since the bus topology consists of only one wire, it is rather inexpensive to implement when compared to other topologies. However, the low cost of implementing the technology is offset by the high cost of managing the network. Additionally, since only one cable is utilized, it can be the single point of failure. If the network cable is terminated on both ends and when without termination data transfer stop and when cable breaks, the entire network will be down.

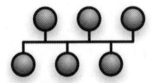

Figure 4-1　Bus Network Topology

2.1　Linear Bus

The type of network topology in which all of the nodes of the network are connected to a common transmission medium which has exactly two endpoints (this is the 'bus', which is also commonly referred to as the backbone, or trunk) – all data that is transmitted between nodes in the network is transmitted over this common transmission medium and is able to be received by all nodes in the network simultaneously.

Note: the two endpoints of the common transmission medium are normally terminated with a device called a terminator[1]. It exhibits the characteristic impedance of the transmission medium and dissipates or absorbs the energy that remains in the signal to prevent the signal from being reflected or propagated back onto the transmission medium in the opposite direction, which would cause interference with and degradation of the signals on the transmission medium.

2.2　Distributed Bus

The type of network topology in which all of the nodes of the network are connected to a common transmission medium which has more than two endpoints that are created by adding branches to the main section of the transmission medium – the physical distributed bus topology functions in exactly the same fashion as the physical linear bus topology (i.e., all nodes share a common transmission medium).

Notes:

- All the endpoints of the common transmission medium are normally terminated using 50 ohm resistor.
- The linear bus topology is sometimes considered to be a special case of the distributed bus topology – i.e., a distributed bus with no branching segments.
- The physical distributed bus topology is sometimes incorrectly referred to as a physical

---

1　Electrical termination of a signal involves providing a terminator at the end of a wire or cable to prevent an RF signal from being reflected back from the end, causing interference. The terminator is placed at the end of a transmission line or daisy chain bus, designed to match impedance and hence minimize signal reflections.

tree topology – however, although the physical distributed bus topology resembles the physical tree topology, it differs from the physical tree topology in that there is no central node to which any other nodes are connected, since this hierarchical functionality is replaced by the common bus.

3. Star

In local area networks with a star topology, each network host is connected to a central hub with a point-to-point connection (see Figure 4-2). In star topology every node (computer workstation or any other peripheral) is connected to the central node called the hub or the switch. The switch is the server and the peripherals are the clients. The network does not necessarily have to resemble a star to be classified as a star network, but all of the nodes on the network must be connected to one central device. All traffic that traverses the network passes through the central hub. The hub acts as a signal repeater[1]. The star topology is considered the easiest topology to design and implement. An advantage of the star topology is the simplicity of adding additional nodes. The primary disadvantage of the star topology is that the hub represents a single point of failure.

Figure 4-2   Star Network

Notes:
- A point-to-point link (described above) is sometimes categorized as a special instance of the physical star topology – therefore, the simplest type of network that is based upon the physical star topology would consist of one node with a single point-to-point link to a second node. It is arbitrary to choose which node is the "hub" and which node is the "spoke".
- After the special case of the point-to-point link, the next simplest type of network that is based upon the physical star topology would consist of one central node – the "hub" – with two separate point-to-point links to two peripheral nodes – the "spokes".
- Although most networks that are based upon the physical star topology are commonly implemented using a special device such as a hub or switch as the central node (i.e., the "hub" of the star), it is also possible to implement a network that is based upon the physical star topology using a computer or even a simple common connection point as the "hub" or central node.
- Star networks may also be described as either broadcast multi-access or nonbroadcast multi-access (NBMA), depending on whether the technology of the network either automatically propagates a signal at the hub to all spokes, or only addresses individual spokes with each communication.

---

1   A repeater is an electronic device that receives a signal and retransmits it at a higher level or higher power, or onto the other side of an <u>obstruction</u>（[əbˈstrʌkʃn] n. 阻塞，障碍物），so that the signal can cover longer distances.

4. Ring

A network topology that is set up in a circular fashion in which data travels around the ring in one direction and each device on the right acts as a repeater to keep the signal strong as it travels (see Figure 4-3). Each device incorporates a receiver for the incoming signal and a transmitter to send the data on to the next device in the ring. The network is dependent on the ability of the signal to travel around the ring.

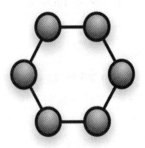

Figure 4-3　Ring Network

5. Tree

The type of network topology in which a central "root" node (the top level of the hierarchy) is connected to one or more other nodes that are one level lower in the hierarchy (i.e., the second level) with a point-to-point link between each of the second level nodes and the top level central "root" node (see Figure 4-4). Each of the second level nodes will also have one or more other nodes that are one level lower in the hierarchy (i.e., the third level) connected to it, also with a point-to-point link, the top level central

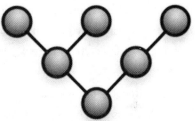

Figure 4-4　Tree Network

"root" node being the only node that has no other node above it in the hierarchy. The hierarchy of the tree is symmetrical. This tree has individual peripheral nodes.

Notes:
- A network that is based upon the physical hierarchical topology must have at least three levels in the hierarchy of the tree, since a network with a central "root" node and only one hierarchical level below it would exhibit the physical topology of a star.
- A network that is based upon the physical hierarchical topology and with a branching factor of 1 would be classified as a physical linear topology.
- The branching factor, f, is independent of the total number of nodes in the network and, therefore, if the nodes in the network require ports for connection to other nodes the total number of ports per node may be kept low even though the total number of nodes is large – this makes the effect of the cost of adding ports to each node totally dependent upon the branching factor and may therefore be kept as low as required without any effect upon the total number of nodes that are possible.
- The total number of point-to-point links in a network that is based upon the physical hierarchical topology will be one less than the total number of nodes in the network.
- If the nodes in a network that is based upon the physical hierarchical topology are required to perform any processing upon the data that is transmitted between nodes in the network, the nodes that are at higher levels in the hierarchy will be required to

perform more processing operations on behalf of other nodes than the nodes that are lower in the hierarchy. Such a type of network topology is very useful and highly recommended.

6. Hybrid

Hybrid networks use a combination of any two or more topologies in such a way that the resulting network does not exhibit one of the standard topologies (e.g., bus, star, ring, etc.). For example, a tree network connected to a tree network is still a tree network topology. A hybrid topology is always produced when two different basic network topologies are connected. Two common examples for hybrid network are star ring network and star bus network.

7. Daisy Chain

Except for star-based networks, the easiest way to add more computers into a network is by daisy-chaining[1], or connecting each computer in series to the next. If a message is intended for a computer partway down the line, each system bounces it along in sequence until it reaches the destination. A daisy-chained network can take two basic forms: linear and ring.

Notes:
- A linear topology puts a two-way link between one computer and the next. However, this was expensive in the early days of computing, since each computer (except for the ones at each end) required two receivers and two transmitters.
- By connecting the computers at each end, a ring topology can be formed. An advantage of the ring is that the number of transmitters and receivers can be cut in half, since a message will eventually loop all of the way around. When a node sends a message, the message is processed by each computer in the ring. If the ring breaks at a particular link then the transmission can be sent via the reverse path, thereby ensuring that all nodes are always connected in the case of a single failure.

New Words

topology	[tə'pɒlədʒi]	n. 拓扑，布局
arrangement	[ə'reɪndʒmənt]	n. 排列，安排
link	[lɪŋk]	n. & vt. 链接
depict	[dɪ'pɪkt]	vt. 描述，描写
placement	['pleɪsmənt]	n. 放置，布置
interconnection	[ˌɪntəkə'nekʃn]	n. 互联
identical	[aɪ'dentɪkl]	adj. 同一的，同样的
telecommunication	[ˌtelɪkəˌmjuːnɪ'keɪʃn]	n. 电信，长途通信，无线电通信，电信学
dynamical	[daɪ'næmɪk]	adj. 动态的

---

1 In electrical and electronic engineering, a daisy chain is a <u>wiring scheme</u>（接线图，布线图）in which multiple devices are wired together in sequence or in a ring. Other than a full, single loop, systems which contain internal loops cannot be called daisy chains.

reconfigure	[ˌriːkən'fɪɡə]	v. 重新装配，改装
permanent	['pɜːmənənt]	adj. 永久的，持久的
endpoint	['ɛnd,pɔɪnt]	n. 端点，终点
unimpeded	[ˌʌnɪm'piːdɪd]	adj. 未受阻止的，没受到阻碍的
proportional	[prə'pɔːʃənl]	adj. 相称的，均衡的
bus	[bʌs]	n. 总线
recipient	[rɪ'sɪpɪənt]	n. 接收者
ignore	[ɪɡ'nɔː]	vt. 不理睬，忽略
utilize	['juːtəlaɪz]	vt. 利用
backbone	['bækbəʊn]	n. 骨干，支柱
trunk	[trʌŋk]	n. 干线，树干，主干
simultaneously	[ˌsɪməl'teɪnɪəsli]	adv. 同时地
terminator	['tɜːmɪneɪtə]	n. 终结器
impedance	[ɪm'piːdns]	n. 阻抗，电阻
dissipate	['dɪsɪpeɪt]	v. 驱散，消耗
propagate	['prɒpəɡeɪt]	v. 传播
resistor	[rɪ'zɪstə]	n. 电阻器
peripheral	[pə'rɪfərəl]	adj. 外围的
		n. 外围设备
hub	[hʌb]	n. 网络集线器，网络中心
spoke	[spəʊk]	n. 轮辐
symmetrical	[sɪ'metrɪkl]	adj. 对称的，均匀的
branching	['brɑːntʃɪŋ]	n. 分歧
		adj. 发枝的
port	[pɔːt]	n. 端口
partway	[ˌpɑːt'weɪ]	v. 到中途，到达一半

Phrases

physical topology	物理拓扑
logical topology	逻辑拓扑
data flow	数据流
regardless of	不管，不顾
transmission rate	传输率
cable layout	电缆配线图，电缆敷设图
act on…	对……起作用，按……行动，作用于
without regard to	不考虑，不遵守
twisted pair Ethernet	双绞线以太网
star topology	星状拓扑

token ring	令牌网
Media Access Unit	媒体存取单元，媒体访问单元
Media Access Control	媒体存取控制，媒体访问控制
Metcalfe's Law	梅特卡夫定律
nail down	钉牢
circuit switching	线路交换，线路转接
packet switching	包交换技术
opposite direction	反向,相反方向
base upon	根据，依据
daisy chain	串式链接，链接式
in sequence	顺次，依次
cut in half	切成两半

Abbreviations

LAN (Local Area Network)	局域网

Exercises

【EX.6】Fill in the blanks according to the text.

1. Network topology is _____ of a computer or biological network.
2. There are two basic categories of network topologies: _____ and _____.
3. Physical topology refers to _____, including device location and cable installation, while logical topology shows _____, regardless of its physical design.
4. The value of a permanent point-to-point network is _____. The value of an on-demand point-to-point connection is _____.
5. Linear bus topology is the type of network topology in which _____ are connected to a common transmission medium which has exactly two endpoints. The two endpoints of the common transmission medium are normally terminated with a device called _____.
6. Distributed bus topology is the type of network topology in which all of the nodes of the network are connected to _____ which has more than two endpoints that are created by adding branches to _____.
7. In star topology every node is connected to the central node called _____. An advantage of the star topology is _____. The primary disadvantage of the star topology is that _____.
8. A network that is based upon the physical hierarchical topology must have at least _____ in the hierarchy of the tree, since a network with a central "root" node and only one hierarchical level below it would exhibit _____.
9. Two common examples for hybrid network are _____ and _____.
10. The two basic forms a daisy-chained network can take are _____ and _____.

【EX.7】 Translate the following terms or phrases from English into Chinese and vice versa.

1. physical topology
2. circuit switching
3. daisy chain
4. packet switching
5. topology
6. placement
7. telecommunication
8. reconfigure
9. backbone
10. proportional
11. *n.* 终结器
12. *n.* 端口
13. *adj.* 外围的 *n.* 外围设备
14. *n.* 网络集线器，网络中心
15. *n.* 终结器

【EX.8】 Translate the following sentences into Chinese.
1. Structural topology optimization method is a newly developing method and has achieved remarkable success.
2. Communication technology actually means the interconnection between equipment to equipment.
3. In a huge telecommunication network, there are many devices from different manufactures.
4. The network transport endpoint already has an address associated with it.
5. Today's Ethernet networks—particularly on the backbone—use switched rather than shared Ethernet.
6. This workstation does not have sufficient resources to open another network session.
7. PCI bus can interconnect peripheral components with CPU.
8. One way to stand out in a crowded category of electronic devices is to employ a novel hardware design.
9. Cable is not symmetrical: downstream speeds far outpace upstream speeds.
10. It is a combination of information technology and management knowledge.

Reading Materials

Internet Terms

1. Internet

The Internet, sometimes called simply "the Net", is a worldwide system of computer networks—a network of networks in which users at any one computer can, if they have permission[1], get information from any other computer (and sometimes talk directly to users at other computers). It was conceived by the Advanced Research Projects Agency (ARPA[2]) of the U.S. government in 1969 and was first known as the ARPANet. The original aim was to create a network that would allow users of a research computer at one university to be able to "talk to" research computers at other universities. A side benefit of ARPANet's design was that, because messages could be routed or rerouted[3] in more than one direction, the network could continue to function even if parts of it were destroyed in the event of a military attack or other disasters.

Today, the Internet is a public, cooperative, and self-sustaining[4] facility accessible to hundreds of millions of people worldwide. Physically, the Internet uses a portion of the total resources of the currently existing public telecommunication networks. Technically, what distinguishes the Internet is its use of a set of protocols called TCP/IP (Transmission Control Protocol/Internet Protocol). Two recent adaptations of Internet technology, the intranet and the extranet, also make use of the TCP/IP protocol.

For many Internet users, electronic mail (e-mail) has practically replaced the Postal Service for short written transactions. Electronic mail is the most widely used application on the Net. You can also carry on live "conversations"[5] with other computer users, using Internet relay chat (IRC[6]). More recently, Internet telephony hardware and software allows real-time voice conversations.

The most widely used part of the Internet is the World Wide Web (often abbreviated "WWW" or called "the Web"). Its outstanding feature is hypertext, a method of instant cross-referencing. In most Web sites, certain words or phrases appear in text of a different color than the rest; often this text is also underlined. When you select one of these words or phrases, you will be transferred to the site or page that is relevant to this word or phrase. Sometimes there are buttons, images, or portions of images that are "clickable"[7]. If you move the pointer over a spot on a Web site and the pointer changes into a hand, this indicates that you can click and be transferred to another site.

---

1 permission [pə(r)'mɪʃ(ə)n] n. 许可，批准，准许
2 Advanced Research Projects Agency (ARPA)：（美国国防部）高级研究计划局
3 reroute [ˌriːˈruːtɪ] v. 改变线路，改道
4 self-sustaining [ˌself səˈsteɪnɪŋ] adj. 自我运行的，自我维持的
5 conversation [ˌkɒnvəˈseɪʃn] n. 交谈，会话
6 Internet relay chat (IRC)：因特网在线聊天
7 clickable [ˈklɪkəbl] adj. 可点击的

Using the Web, you have access to millions of pages of information. Web browsing is done with a Web browser, the most popular of which are Microsoft Internet Explorer and Netscape Navigator. The appearance of a particular Web site may vary slightly depending on the browser you use. Also, later versions of a particular browser are able to render more "bells and whistles[1]" such as animation, virtual reality, sound, and music files, than earlier versions.

2. TCP/IP (Transmission Control Protocol/Internet Protocol)

TCP/IP (Transmission Control Protocol/Internet Protocol) is the basic communication language or protocol of the Internet. It can also be used as a communications protocol in a private network (either an intranet or an extranet). When you are set up with direct access to the Internet, your computer is provided with a copy of the TCP/IP program just as every other computer that you may send messages to or get information from also has a copy of TCP/IP.

TCP/IP is a two-layer program. The higher layer, transmission control protocol, manages the assembling of a message or file into smaller packets that are transmitted over the Internet and received by a TCP layer that reassembles the packets into the original message. The lower layer, internet protocol, handles the address part of each packet so that it gets to the right destination. Each gateway computer on the network checks this address to see where to forward the message. Even though some packets from the same message are routed differently than others, they'll be reassembled at the destination.

TCP/IP uses the client/server model of communication in which a computer user (a client) requests and is provided a service (such as sending a Web page) by another computer (a server) in the network. TCP/IP communication is primarily point-to-point, meaning each communication is from one point or host computer[2] in the network to another point or host computer. TCP/IP and the higher-level applications that use it are collectively said to be "stateless" because each client request is considered a new request unrelated to any previous one (unlike ordinary phone conversations that require a dedicated connection for the call duration[3]). Being stateless frees network paths so that everyone can use them continuously. Note that the TCP layer itself is not stateless as far as any one message is concerned. Its connection remains in place until all packets in a message have been received.

Many Internet users are familiar with the even higher layer application protocols that use TCP/IP to get to the Internet. These include the world wide Web's hypertext transfer protocol (HTTP), the file transfer protocol (FTP), telnet which lets you logon to remote computers, and the simple mail transfer protocol (SMTP[4]). These and other protocols are often packaged together with TCP/IP as a "suite[5]".

---

1 bells and whistles: 花哨
2 host computer: 主机
3 for the duration: 在整段时期内
4 simple mail transfer protocol (SMTP): 简单邮件传送协议
5 suite [swi:t] n. 套件，套，组

Personal computer users with an analog phone modem connection to the Internet usually get to the Internet through the serial line internet protocol (SLIP[1]) or the point-to-point protocol (PPP). These protocols encapsulate the IP packets so that they can be sent over the dial-up phone connection to an access provider's modem.

Protocols related to TCP/IP include the user datagram protocol (UDP[2]), which is used instead of TCP for special purposes. Other protocols are used by network host computers for exchanging router information. These include the internet control message protocol (ICMP[3]), the interior gateway protocol (IGP[4]), the exterior gateway protocol (EGP[5]), and the border gateway protocol (BGP[6]).

3. HTTP (Hypertext Transfer Protocol)

HTTP is the set of rules for transferring files (text, graphic images, sound, video, and other multimedia files) on the world wide Web. As soon as a Web user opens their Web browser, the user is indirectly making use of HTTP. HTTP is an application protocol that runs on top of the TCP/IP suite of protocols (the foundation protocols for the Internet).

HTTP concepts include (as the hypertext part of the name implies) the idea that files can contain references to other files whose selection will elicit additional transfer requests. Any Web server machine contains, in addition to the Web page files it can serve, an HTTP daemon[7], a program that is designed to wait for HTTP requests and handle them when they arrive. Your Web browser is an HTTP client, sending requests to server machines. When the browser user enters file requests by either "opening" a Web file (typing in a uniform resource locator or URL) or clicking on a hypertext link[8], the browser builds an HTTP request and sends it to the internet protocol address (IP address) indicated by the URL. The HTTP daemon in the destination server machine receives the request and sends back the requested file or files associated with the request.

参考译文

Text A 网络体系结构

网络体系结构就是设计通信网络。它是规范网络物理部件、其功能的组织和配置、运行法则和过程及运行中所用的数据格式的框架。

在通信学科中,网络体系结构的规范可能还包括详细描述通过通信网络交付的产品和

1　serial line internet protocol (SLIP):串行线路网际协议
2　user datagram protocol(UDP):用户数据报协议
3　Internet contral message protocol (ICMP):因特网控制消息协议
4　interior gateway protocol (IGP):内部网关协议
5　exterior gateway protocol (EGP):外部网关协议
6　border gateway protocol (BGP):边界网关协议
7　daemon ['di:mən] n. 精灵
8　hypertext link:超链接

服务及此服务详细的收费构成和费率。

1. OSI 网络模型

开放系统互连参考模型（OSI 模型）是由国际标准化组织提出的开放系统互连产品。它提供了一种把通信系统细分为叫作"层"的更小部分的方法。层是相似功能的集合，它为其上的层提供服务并从其下的层接受服务。在每一层，一个实例为其上层实例提供服务并请求下层提供服务。

1.1 物理层

物理层定义设备的电子和物理规范。尤其是它定义了设备和传输介质（如铜线或光缆）之间的关系，包括针的布局、电压、电缆规范、集线器、中继器、网络适配器、主机总线适配器（HBA 用于存储区域网络）等。物理层的主要任务是通过通信通道传输比特流。

1.2 数据链路层

数据链路层提供在网络实体之间传输数据的功能和方法，并检查和尽量改正物理层可能出现的错误。起初，该层打算供点对点和点对多点介质使用，这是电话系统中广域介质的特点。局域网体系包括可广播的多路访问介质，其开发不依赖 IEEE Project 802 中的 ISO。在当今的实践中，在数据链路协议（如点对点协议）只有错误检测而没有使用滑动窗口的流控制，并且，在局域网中，IEEE 802.2 LLC 层在以太网上没有使用太多的协议，在其他局域网中，它的流控制和确认机制也很少使用。滑动窗口流控制和确认由像 TCP 这样的协议用在传输层，但仍然用于 X.25 能够提供性能优势的情况下。简单地说，其主要任务是建立并识别帧边界。可以通过在帧的开头和结尾附加特殊的位格式来实现该任务，这样输入数据就划分为帧。

1.3 网络层

网络层提供在不同网络中从源主机到目的主机之间传输可变长度数据序列的功能和方法，期间保持传输层要求的服务质量（与连接同一网络内主机的数据链路层对应）。网络层执行网络路由功能，也可实现信息的分解与重新装配，并报告传输错误。路由器运行在这一层——通过扩展的网络发送数据并使因特网成为可能。这是一个逻辑寻址方案——值由网络工程师选择。这个寻址方案不分层，它控制子层的运行、确定 IMP 之间的路由策略并确保所有的数据包都在目的地按照适当的顺序正确接收。

1.4 传输层

传输层提供端用户之间数据的透明传输，这样就可以把数据可靠地传输给上层。传输层管理某个流控制链路、分解/合并及错误控制的可靠性。有些协议是面向状态和连接的，这就意味着传输层可以跟踪信息段并在传输失败后重发。传输层也提供对数据传输成功的确认，如果没有错误再发送之后的数据。某些协议（如 TCP 而不是 UDP）支持虚拟线路并通过下层的面向数据包的数据报网络来提供面向连接的通信。数据报传输随机发送数据包并向多个节点广播。

注意：传输层可以把多个流放入一个物理通道中。传输的头信息可以告知消息属于哪个连接。

1.5 会话层

该层提供用户接口以便用户可以建立会话连接。用户必须提供要联系的远程地址。在两个进程间建立会话的操作叫作"绑定"。在有些协议中，它被合并到传输层。其主要工作是从其他应用中向本应用传输数据，因此本应用主要用于传输层。

1.6 表示层

表示层建立应用层实体之间的语境关联，其中，如果表示服务提供相应的映射，则高层实体可用不同的语法和语义。如果映射有效，可以把表示服务数据单元封装到会话协议数据单元，并传递到堆栈。该层通过在应用与网络格式之间转换而独立于数据表示（如加密）。表示层把数据转换为应用可以接受的形式。该层格式化和加密网络上要发送的数据，它有时也叫作语法层。最初的表示结构使用抽象语法表示法1（ASN.1）的基本编码规则，它能够把 EBCDIC 编码的文本文件转换为 ASCII 编码的文件，或者把对象和来自与发往 XML 的其他数据结构序列化。

1.7 应用层

应用层是离端用户最近的 OSI 层。这意味着 OSI 应用层和用户与软件应用程序直接交互。应用层与实现通信的软件应用程序交互。这样的应用程序超出了 OSI 模型的范围。应用层的功能通常包括识别通信伙伴、确定可用资源及同步通信。当识别通信伙伴时，应用层确定通信伙伴的身份和有效性，以便传输数据。

2. 分布计算

在分布计算的独特应用中，术语"网络体系结构"通常用来描述分布应用体系的结构和种类，因为分布式应用中的参与节点通常被称为网络。例如，公用电话交换网络的应用体系被称为"高级智能网"。有许多特殊种类，但都依赖哑网络（如因特网）与智能计算机网络（如电话网）之间的通信。其他网络包含着两类网络的各种要素，以便满足各种应用。最近，情景感知网络——这两者的综合——能够结合两者中的最佳元素，已经获得了人们的更多关注。

在分布应用和 PVC（永久虚电路）中，该术语的一个流行范例是点对点服务和网络中的节点组织。P2P 网络通常实现一个覆盖网络，运行在更低的物理或逻辑网络上。这些覆盖网络可以按照几种明确的模型（系统的网络体系）实现一些确定的节点组织结构。

Unit 5

Text A

Networking Hardware

Networking hardware includes all computers, peripherals, interface cards and other equipment needed to perform data processing and communications within the network (see Figure 5-1).

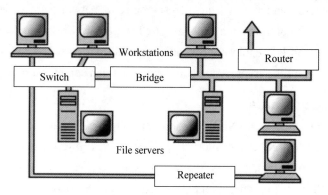

Figure 5-1　Networking Hardware

This section provides information on the following components:
- File servers.
- Workstations.
- Network interface cards.
- Concentrators/hubs.
- Repeaters.
- Bridges.
- Routers.

1. File Servers

A file server stands at the heart of most networks. It is a very fast computer with a large amount of RAM and storage space, along with a fast network interface card. The network operating system software resides on this computer, along with any software applications and data files that need to be shared.

The file server controls the communication of information between the nodes on a network. For example, it may be asked to send a word processor program to one workstation, receive a database file from another workstation, and store an e-mail message during the same time period. This requires a computer that can store a lot of information and share it very quickly. File servers should have at least the following characteristics:

- 75 megahertz or faster microprocessor.
- A fast hard drive with at least four gigabytes of storage.
- A RAID (Redundant Array of Inexpensive Disks) to preserve data after a disk casualty.
- A tape back-up unit.
- Numerous expansion slots.
- Fast network interface card.
- At least of 32 MB of RAM.

2. Workstations

All the computers connected to the file server on a network are called workstations. A typical workstation is a computer that is configured with a network interface card, networking software, and the appropriate cables. Workstations do not necessarily need hard drives because files can be saved on the file server. Almost any computer can serve as a network workstation.

3. Network Interface Cards

The network interface card (NIC) provides the physical connection between the network and the workstation. Most NICs are internal, with the card fitting into an expansion slot inside the computer. Some computers, such as Mac Classics, use external boxes which are attached to a serial port or a SCSI port. Laptop computers generally use external LAN adapters connected to the parallel port or network cards that slip into a PCMCIA[1] slot.

Network interface cards are a major factor in determining the speed and performance of a network. It is a good idea to use the fastest network card available for the type of workstation you are using.

The three most common network interface connections are Ethernet cards, LocalTalk connectors, and token ring cards. According to an International Data Corporation, Ethernet is the most popular, followed by token ring and LocalTalk.

---

1 Short for Personal Computer Memory Card International Association, and pronounced as separate letters, PCMCIA is an organization consisting of some 500 companies that has developed a standard for small, credit card-sized devices, called PC Cards. Originally designed for adding memory to portable (['pɔːtəbl] *adj.* 轻便的，手提（式）的) computers, the PCMCIA standard has been expanded several times and is now suitable for many types of devices.

3.1 Ethernet Cards

Ethernet cards are usually purchased separately from a computer, although many computers (such as the Macintosh) now include an option for a preinstalled Ethernet card. Ethernet cards contain connections for either coaxial or twisted pair cables (or both). If it is designed for coaxial cable, the connection will be BNC[1]. If it is designed for twisted pair, it will have a RJ-45 connection. Some Ethernet cards also contain an AUI[2] connector. This can be used to attach coaxial, twisted pair, or fibre optics cable to an Ethernet card. When this method is used there is always an external transceiver attached to the workstation.

3.2 Token Ring Cards

Token ring network cards look similar to Ethernet cards. One visible difference is the type of connector on the back end of the card. Token ring cards generally have a nine pin DIN type connector to attach the card to the network cable.

4. Concentrators/Hubs

A concentrator is a device that provides a central connection point for cables from workstations, servers, and peripherals. In a star topology, twisted-pair wire is run from each workstation to a central concentrator. Hubs are multi-slot concentrators into which a number of multi-port cards can be plugged to provide additional access as the network grows in size. Some concentrators are passive. They allow the signal to pass from one computer to another without any change. Most concentrators are active. They electrically amplify the signal as it moves from one device to another. Active concentrators, like repeaters, are used to extend the length of a network. Concentrators are:

- Usually configured with 8, 12, or 24 RJ-45 ports.
- Often used in a star or star-wired ring topology.
- Sold with specialized software for port management.
- Also called hubs.
- Usually installed in a standardized metal rack that also may store netmodems, bridges, or routers.

5. Repeaters

When a signal travels along a cable, it tends to lose strength. A repeater is a device that boosts a network's signal as it passes through. The repeater does this by electrically amplifying the signal it receives and rebroadcasting it. Repeaters can be separate devices or they can be

---

1 The BNC connector is a miniature quick connect/disconnect RF connector used for coaxial cable. It features two bayonet lugs on the female connector; mating is achieved with only a quarter turn of the <u>coupling nut</u>（连接螺母）. BNCs are ideally suited for cable termination for miniature-to-<u>subminiature</u> ([sʌb'minjətʃə] *adj.* 超小型的，微型的) coaxial cable (e.g., RG-58, 59, to RG-179, RG-316).

2 An attachment unit interface (AUI) is a 15 pin connection that provides a path between a node's Ethernet interface and the <u>medium attachment unit</u>（MAU，介质连接单元），sometimes known as a transceiver. It is the part of the IEEE Ethernet standard located between the media access control (MAC), and the MAU. An AUI cable may be up to 50 meters long, although frequently the cable is omitted altogether and the MAU and MAC are directly attached to one another.

incorporated into a concentrator. They are used when the total length of your network cable exceeds the standards set for the type of cable being used.

A good example of the use of repeaters would be in a local area network using a star topology with unshielded twisted-pair cabling. The length limit for unshielded twisted-pair cable is 100 meters. The most common configuration is for each workstation to be connected by twisted-pair cable to a multi-port active concentrator. The concentrator regenerates all the signals that pass through it allowing for the total length of cable on the network to exceed the 100 meter limit.

6. Bridges

A bridge is a device that allows you to segment a large network into two smaller, more efficient networks. If you are adding to an older wiring scheme and want the new network to be up-to-date, a bridge can connect the two.

A bridge monitors the information traffic on both sides of the network so that it can pass packets of information to the correct location. Most bridges can "listen" to the network and automatically figure out the address of each computer on both sides of the bridge. The bridge can inspect each message and, if necessary, broadcast it on the other side of the network.

The bridge manages the traffic to maintain optimum performance on both sides of the network. You might say that the bridge is like a traffic cop at a busy intersection during rush hour. It keeps information flowing on both sides of the network, but it does not allow unnecessary traffic through. Bridges can be used to connect different types of cabling, or physical topologies. They must, however, be used between networks with the same protocol.

7. Routers

A router translates information from one network to another; it is similar to a super intelligent bridge. Routers select the best path to route a message, based on the destination address and origin. The router can direct traffic to prevent head-on collisions, and is smart enough to know when to direct traffic along back roads and shortcuts.

While bridges know the addresses of all computers on each side of the network, routers know the addresses of computers, bridges, and other routers on the network. Routers can even "listen" to the entire network to determine which sections are busiest—they can then redirect data around those sections until they clear up.

If you have a school LAN that you want to connect to the Internet, you will need to purchase a router. In this case, the router serves as the translator between the information on your LAN and the Internet. It also determines the best route to send the data over the Internet. Routers can:

- Direct signal traffic efficiently.
- Route messages between any two protocols.

- Route messages between linear bus, star, and star-wired ring[1] topologies.
- Route messages across fiber optic, coaxial, and twisted-pair cabling.

New Words

concentrator	['kɒnsəntreɪtə]	n. 集中器
bridge	[brɪdʒ]	n. 桥接器
reside	[rɪ'zaɪd]	vi. 驻留
share	[ʃeə]	n. & v. 分享，共享
workstation	['wɜːksteɪʃn]	n. 工作站
megahertz	['megəhɜːts]	n. 兆赫
microprocessor	[ˌmaɪkrəʊ'prəʊsesə]	n. 微处理器
gigabyte	['gɪgəbaɪt]	n. 十亿字节，吉字节
preserve	[prɪ'zɜːv]	vt. 保护，保持，保存
casualty	['kæʒuəlti]	n. 损坏，事故
configure	[kən'fɪgə]	vi. 配置，设定
internal	[ɪn'tɜːnl]	adj. 内在的，内部的
external	[ɪk'stɜːnl]	adj. 外部的
determine	[dɪ'tɜːmɪn]	vt. 决定，断定
option	['ɒpʃn]	n. 选项，选择权
preinstall	['priːɪn'stɔːl]	v. 预设，预安装
coaxial	['kəʊ'æksɪəl]	adj. 同轴的，共轴的
transceiver	[træn'siːvə]	n. 收发器
connector	[kə'nektə]	n. 连接器
multislot	['mʌltɪslɒt]	n. 多插槽，多插座
passive	['pæsɪv]	adj. 被动的
active	['æktɪv]	adj. 主动的，活动的
amplify	['æmplɪfaɪ]	vt. 放大，增强
standardized	['stændədaɪzd]	adj. 标准的
rack	[ræk]	n. 架，设备架 vt. 放在架上
netmodem	[net'məʊdem]	n. 网络调制解调器
boost	[buːst]	v. 推进
rebroadcast	[riː'brɔːdkɑːst]	v. & n. 转播，重播
exceed	[ɪk'siːd]	vt. 超越，胜过 vi. 超过其他

---

1 A star-wired ring topology may appear (externally) to be the same as a star topology. Internally, the <u>MAU</u> (multistation access unit,多站访问部件) of a star-wired ring contains wiring that allows information to pass from one device to another in a circle or ring. The token ring protocol uses a star-wired ring topology.

unshielded	[ʌn'ʃiːldɪd]	adj. 无防护的，无铠装的，无屏蔽的
regenerate	[rɪ'dʒenəreɪt]	vt. 使新生，重建
segment	[seg'ment]	v. 分割
	['segmənt]	n. 段，节，片断
address	[ə'dres]	n. 地址
inspect	[ɪn'spekt]	v. 检查
optimum	['ɒptɪməm]	n. 最适宜
		adj. 最佳的
intersection	[ˌɪntə'sekʃn]	n. 十字路口
unnecessary	[ʌn'nesəsərɪ]	adj. 不必要的，多余的
intelligent	[ɪn'telɪdʒənt]	adj. 聪明的，智能的
route	[ruːt]	v. 发送
		n. 路线，路程，通道
collision	[kə'lɪʒn]	n. 碰撞，冲突
shortcut	['ʃɔːtkʌt]	n. 捷径
redirect	[ˌriːdə'rekt]	vt. 重寄，使改道，使改变方向

Phrases

interface card	接口卡
network interface card (NIC)	网络接口卡
file server	文件服务器
storage space	存储空间
along with…	连同……一起，随同……一起
network operating system	网络操作系统
software application	软件应用程序
data file	数据文件
word processor	文字处理软件
database file	数据库文件
hard drive	硬盘驱动器
tape back-up unit	磁带备份机
be saved on…	被保存在……上
fit into	插入，装入
attach… to	附在……
serial port	串行端口
laptop computer	膝上型计算机
parallel port	并行端口
network card	网卡
slip into	分成

twisted pair	双绞线
fibre optics cable	光导纤维电缆
port management	端口管理
be incorporated into	融入
unshielded twisted-pair	非屏蔽双绞线
information traffic	信息流量
on both sides	双方，两边
figure out	计算出，断定
traffic cop	<美口>交通警察
rush hour	高峰时间
be similar to…	与……相似
clear up	整理，消除

Abbreviations

RAM (Random Access Memory)	随机存储器
RAID (Redundant Array of Inexpensive Disks)	独立磁盘冗余阵列
MB (Megabyte)	兆字节
SCSI (Small Computer System Interface)	小型计算机系统接口
PCMCIA (Personal Computer Memory Card International Association)	个人计算机存储卡国际联盟
BNC (Bayonet Nut Connector)	同轴电缆接插件
AUI (Attachment Unit Interface)	连接单元接口
DIN (Deutsche Industrie-Norm (德文))	德国工业标准

Analysis of Difficult Sentences

[1] Networking hardware includes all computers, peripherals, interface cards and other equipment needed to perform data processing and communications within the network.

本句中，needed to perform data processing and communications within the network 是一个过去分词短语，作定语，修饰和限定 other equipment。它可以扩展为一个定语从句：which are needed to perform data processing and communications within the network。

[2] A typical workstation is a computer that is configured with a network interface card, networking software, and the appropriate cables.

本句中，that is configured with a network interface card, networking software, and the appropriate cables 是一个定语从句，修饰和限定 a computer。

[3] Laptop computers generally use external LAN adapters connected to the parallel port or network cards that slip into a PCMCIAslot.

本句中，connected to the parallel port 是一个过去分词短语，作定语，修饰和限定 external LAN adapters。that slip into a PCMCIAslot 是一个定语从句，修饰和限定 network cards。

[4] Hubs are multi-slot concentrators into which a number of multi-port cards can be plugged to provide additional access as the network grows in size.

本句中，into which a number of multi-port cards can be plugged to provide additional access as the network grows in size 是一个介词前置的定语从句，修饰和限定 multi-slot concentrators。在该从句中，to provide additional access 作目的状语，as the network grows in size 作时间状语。

[5] While bridges know the addresses of all computers on each side of the network, routers know the addresses of computers, bridges, and other routers on the network.

本句中，While bridges know the addresses of all computers on each side of the network 是一个让步状语从句，while 的意思是"尽管，虽然"，等于 although。

Exercises

【EX.1】 Answer the following questions according to the text.

1. What does networking hardware include?
2. What is a file server?
3. What does a file server do?
4. What is a typical workstation?
5. What does the network interface card (NIC) do?
6. What are the three most common network interface connections?
7. What is a concentrator?
8. How does a repeater boost a network's signal as it passes through?
9. What does a bridge do?
10. If you have a school LAN that you want to connect to the Internet, what will you need to buy?

【EX.2】 Translate the following terms or phrases from English into Chinese and vice versa.

1.	network operating system	1.	
2.	serial port	2.	
3.	hard drive	3.	
4.	fiber optics cable	4.	
5.	unshielded twisted-pair	5.	
6.	parallel port	6.	
7.	bridge	7.	
8.	microprocessor	8.	
9.	preinstall	9.	
10.	connector	10.	
11.	*n.* 收发器	11.	

12.	*n.* 地址	12.	
13.	*n.* 碰撞，冲突	13.	
14.	*v.* 分割　*n.* 段，节，片断	14.	
15.	*vi.* 配置，设定	15.	

【EX.3】 **Translate the following sentences into Chinese.**

1. Data transmit between the concentrator and the collector in the form of wireless ad hoc networks.
2. Normally demodulation frequency is inside the limits of a few hertz to hundreds of megahertz.
3. The second option is technically superior but it demands higher-performance equipment.
4. Coaxial cable is also used for undersea telephone lines.
5. We are using this transistor to amplify a telephone signal.
6. It can regenerate data in storage units where the process of reading data results in its destruction.
7. Please inspect all parts for damage prior to installation and start-up.
8. Microsoft developed an intelligent solution to this problem.
9. What function would you use to redirect the browser to a new page?
10. Filter route establishes the basic rules for connectivity through a firewall.

【EX.4】 **Complete the following passage with appropriate words in the box.**

provides	devices	host	surrounded	signals
carrier	random	amplify	electricity	designed

　　CSMA/CA stands for carrier sense multiple access collision avoidance. It is a network access method in which each device signals its intent to transmit before it actually does so. This prevents other ___1___ from sending information, thus preventing collisions from occurring between ___2___ from two or more devices. This is the access method used by LocalTalk.

　　CSMA/CD stands for carrier sense multiple access collision detection. It is a network access method in which devices that are ready to transmit data first check the channel for a carrier. If no ___3___ is sensed, a device can transmit. If two devices transmit at once, a collision occurs and each computer backs off and waits a ___4___ amount of time before attempting to retransmit. This is the access method used by Ethernet.

　　Concentrator is a device that ___5___ a central connection point for cables from workstations, servers, and peripherals. Most concentrators contain the ability to ___6___ the electrical signal they receive.

　　Dumb terminal refers to devices that are ___7___ to communicate exclusively with a host (main frame) computer. It receives all screen layouts from the host computer and sends all keyboard entry to the host. It cannot function without the ___8___ computer.

Fibre optic cable is a a cable which consists of a center glass core ___9___ by layers of plastic. It transmits data using light rather than ___10___. It has the ability to carry more information over much longer distances.

【EX.5】 Translate the following passage into Chinese.

Network Gateway

A network gateway is an internetworking system capable of joining together two networks that use different base protocols. A network gateway can be implemented completely in software, completely in hardware, or as a combination of both. Depending on the types of protocols they support, network gateways can operate at any level of the OSI model.

Because a network gateway, by definition, appears at the edge of a network, related capabilities like firewalls tend to be integrated with it. On home networks, a broadband router typically serves as the network gateway although ordinary computers can also be configured to perform equivalent functions.

Text B

Network Switch

A network switch is a computer networking device that links network segments or network devices. The term commonly refers to a multi-port network bridge that processes and routes data at the data link layer (layer 2) of the OSI model. Switches that additionally process data at the network layer (layer 3) and above are often called layer-3 switches or multilayer switches.

1. Function

A switch is a telecommunication device which receives a message from any device connected to it and then transmits the message only to the device for which the message is meant. This makes the switch a more intelligent device than a hub (which receives a message and then transmits it to all the other devices on its network). The network switch plays an integral part in most modern Ethernet local area networks (LANs). Mid-to-large sized LANs contain a number of linked managed switches. Small office/home office (SOHO) applications typically use a single switch, or an all-purpose converged device such as a residential gateway to access small office/home broadband services such as DSL or cable Internet. In most cases, the end-user device contains a router and components that interface to the particular physical broadband technology. User devices may also include a telephone interface for VoIP[1].

An Ethernet switch operates at the data link layer of the OSI model to create a separate

---

1 Voice over IP (VoIP, abbreviation of voice over internet protocol) commonly refers to the communication protocols, technologies, methodologies, and transmission techniques involved in the delivery of voice communications and multimedia sessions over internet protocol (IP) networks, such as the Internet.

collision domain[1] for each switch port. With 4 computers (e.g., A, B, C, and D) on 4 switch ports, any pair (e.g. A and B) can transfer data back and forth while the other pair (e.g. C and D) also do so simultaneously, and the two conversations will not interfere with one another. In full duplex[2] mode, these pairs can also overlap (e.g. A transmits to B, simultaneously B to C, and so on). In the case of a repeater hub, they will all share the bandwidth and run in half duplex[3], resulting in collisions, which will then necessitate retransmissions.

Using a bridge or a switch (or a router) to split a larger collision domain into smaller ones in order to reduce collision probability and improve overall throughput is called segmentation. In the extreme of microsegmentation, each device is located on a dedicated switch port. In contrast to an Ethernet hub, there is a separate collision domain on each of the switch ports. This allows computers to have dedicated bandwidth on point-to-point connections to the network and also to run in full duplex without collisions. Full duplex mode has only one transmitter and one receiver per "collision domain", making collisions impossible.

2. Role of Switches in a Network

Switches may operate at one or more layers of the OSI model, including data link and network. A device that operates simultaneously at more than one of these layers is known as a multilayer switch.

In switches intended for commercial use, built-in or modular interfaces make it possible to connect different types of networks, including Ethernet, fibre channel, ATM, ITU-T[4]G.hn[5] and 802.11[6]. This connectivity can be at any of the layers mentioned. While layer-2 functionality is adequate for bandwidth-shifting within one technology, interconnecting technologies such as Ethernet and token ring is easier at layer 3.

Devices that interconnect at layer 3 are traditionally called routers, so layer-3 switches can

---

1 A collision domain is a section of a network where data packets can collide with（冲突）one another when being sent on a shared medium or through repeaters, in particular, when using early versions of Ethernet. A network collision occurs when more than one device attempts to send a packet on a network segment at the same time. Collisions are resolved using carrier sense multiple access （多路存取）with collision detection in which the competing packets are discarded and resent one at a time. This becomes a source of inefficiency（[ˌɪnɪˈfɪʃənsi] n. 无效率，无能）in the network.

2 A full-duplex (FDX), or sometimes double-duplex system, allows communication in both directions, and, unlike half-duplex, allows this to happen simultaneously. Land-line（陆线） telephone networks are full-duplex, since they allow both callers to speak and be heard at the same time.

3 A half-duplex (HDX) system provides communication in both directions, but only one direction at a time (not simultaneously). Typically, once a party begins receiving a signal, it must wait for the transmitter to stop transmitting, before replying (antennas are of trans-receiver type in these devices, so as to transmit and receive the signal as well).

4 The ITU Telecommunication Standardization Sector (ITU-T) is one of the three sectors (divisions or units) of the International Telecommunication Union (ITU); it coordinates standards for telecommunications.

5 G.hn is the common name for a home network technology family of standards developed under the International Telecommunication Union's Telecommunication Standardization Sector (the ITU-T) and promoted by the HomeGrid Forum and several other organizations. The G.hn specification defines networking over power lines（电力线）, phone lines and coaxial cables with data rates up to 1Gbit/s.

6 IEEE 802.11 is a set of standards for implementing wireless local area network（WLAN, 无线局域网）computer communication in the 2.4, 3.6 and 5 GHz frequency bands（频段）.

also be regarded as (relatively primitive) routers.

Where there is a need for a great deal of analysis of network performance and security, switches may be connected between WAN routers as places for analytic modules. Some vendors provide firewall, network intrusion detection, and performance analysis modules that can plug into switch ports. Some of these functions may be on combined modules.

In other cases, the switch is used to create a mirror image of data that can go to an external device. Since most switch port mirroring provides only one mirrored stream, network hubs can be useful for fanning out data to several read-only analyzers, such as intrusion detection systems and packet sniffers.

3. Layer-specific Functionality

While switches may learn about topologies at many layers, and forward at one or more layers, they do tend to have common features. Other than for high performance applications, modern commercial switches use primarily Ethernet interfaces.

At any layer, a modern switch may implement power over Ethernet (PoE), which avoids the need for attached devices, such as a VoIP phone or wireless access point, to have a separate power supply. Since switches can have redundant power circuits connected to uninterruptible power supplies, the connected device can continue operating even when regular office power fails.

3.1 Layer 1(Hubs Versus Higher-layer Switches)

A network hub, or repeater, is a simple network device. Hubs do not manage any of the traffic that comes through them. Any packet entering a port is broadcast out or "repeated" on every other port, except for the port of entry. Since every packet is repeated on every other port, packet collisions affect the entire network, limiting its capacity.

A switch creates the—originally mandatory—layer 1 end-to-end connection only virtually. Its bridge function selects which packets are forwarded to which port(s). The connection lines are not "switched" literally, it only appears like this on the packet level. "bridging hub" or possibly "switching hub" would be more appropriate terms.

There are specialized applications where a hub can be useful, such as copying traffic to multiple network sensors. High end switches have a feature which does the same thing called port mirroring[1].

3.2 Layer 2

A network bridge, operating at the data link layer, may interconnect a small number of devices in a home or the office. This is a trivial case of bridging, in which the bridge learns the MAC address of each connected device.

Single bridges can also provide extremely high performance in specialized applications

---

1 Port mirroring is used on a network switch to send a copy of network packets seen on one switch port (or an entire VLAN) to a network monitoring connection on another switch port. This is commonly used for network appliances that require monitoring of network traffic, such as an intrusion detection system.

such as storage area networks.

Classic bridges may also interconnect using a spanning tree protocol[1] that disables links so that the resulting local area network is a tree without loops. In contrast to routers, spanning tree bridges must have topologies with only one active path between two points. The older IEEE 802.1D spanning tree protocol could be quite slow, with forwarding stopping for 30 seconds while the spanning tree would reconverge. A rapid spanning tree protocol was introduced as IEEE 802.1w. The newest standard shortest path Bridging (IEEE 802.1aq) is the next logical progression and incorporates all the older spanning tree protocols (IEEE 802.1D STP, IEEE 802.1w RSTP, IEEE 802.1s MSTP) that blocked traffic on all but one alternative path. IEEE 802.1aq (Shortest Path Bridging, SPB) allows all paths to be active with multiple equal cost paths, provides much larger layer 2 topologies (up to 16 million compared to the 4096 VLANs limit), faster convergence times, and improves the use of the mesh topologies through increase bandwidth and redundancy between all devices by allowing traffic to load share across all paths of a mesh network.

While layer 2 switch remains more of a marketing term than a technical term, the products that were introduced as "switches" tended to use microsegmentation and full duplex to prevent collisions among devices connected to Ethernet. By using an internal forwarding plane which is much faster than any interface, they give the impression of simultaneous paths among multiple devices. "Non-blocking" devices use a forwarding plane or equivalent method which is fast enough to allow full duplex traffic for each port simultaneously.

Once a bridge learns the addresses of its connected nodes, it forwards data link layer frames using a layer 2 forwarding method. There are four forwarding methods a bridge can use, of which the second through fourth method were performance-increasing methods when used on "switch" products with the same input and output port bandwidths:

- Store and forward: the switch buffers and verifies each frame before forwarding it.
- Cut through: the switch reads only up to the frame's hardware address before starting to forward it. Cut-through switches have to fall back to store and forward if the outgoing port is busy at the time the packet arrives. There is no error checking with this method.
- Fragment free: a method that attempts to retain the benefits of both store and forward and cut through. Fragment free checks the first 64 bytes of the frame, where addressing information is stored. According to Ethernet specifications, collisions should be detected during the first 64 bytes of the frame, so frames that are in error because of a collision will not be forwarded. This way the frame will always reach its intended destination. Error checking of the actual data in the packet is left for the end device.
- Adaptive switching: a method of automatically selecting between the other three modes.

---

1 The spanning tree protocol (STP) is a network protocol that ensures a loop-free topology for any bridged Ethernet local area network. The basic function of STP is to prevent bridge loops and the broadcast <u>radiation</u>（[ˌreɪdɪˈeɪʃn] n. 辐射，放射）that results from them.

While there are specialized applications, such as storage area networks, where the input and output interfaces are of the same bandwidth, this is not always the case in general LAN applications. In LANs, a switch used for end user access typically concentrates lower bandwidth and uplinks into a higher bandwidth.

3.3 Layer 3

Within the confines of the Ethernet physical layer, a layer 3 switch can perform some or all of the functions normally performed by a router. The most common layer 3 capability is the awareness of IP multicast[1] through IGMP snooping[2]. With this awareness, a layer 3 switch can increase efficiency by delivering the traffic of a multicast group only to ports where the attached device has signaled that it wants to listen to that group.

3.4 Layer 4

While the exact meaning of the term layer 4 switch is vendor-dependent, it almost always starts with a capability for network address translation, but then adds some type of load distribution based on TCP sessions.

The device may include a stateful inspection firewall[3], a VPN concentrator, or be an IPSec[4] security gateway.

3.5 Layer 7

Layer 7 switches may distribute loads based on URL(Uniform Resource Locator) or by some installation-specific technique to recognize application-level transactions. A layer 7 switch may include a web cache and participate in a content delivery network[5].

4. Types of Switches

4.1 Form Factor

- Desktop, not mounted in an enclosure, typically intended to be used in a home or office

---

1　IP multicast is a method of sending internet protocol (IP) datagrams to a group of interested receivers in a single transmission. It is often employed for streaming media applications on the Internet and private networks. The method is the IP-specific version of the general concept of multicast networking. It uses specially reserved multicast address blocks in IPv4 and IPv6.

2　IGMP snooping is the process of listening to internet group management protocol (IGMP) network traffic. The feature allows a network switch to listen in on the IGMP conversation between hosts and routers. By listening to these conversations the switch maintains a map of which links need which IP multicast streams.

3　A firewall can either be software-based or hardware-based and is used to help keep a network secure. Its primary objective is to control the incoming and outgoing network traffic by analyzing the data packets and determining whether it should be allowed through or not, based on a predetermined rule set（规则集）.

4　Internet protocol security (IPSec) is a protocol suite for securing internet protocol (IP) communications by authenticating and encrypting each IP packet of a communication session. IPsec also includes protocols for establishing mutual（['mjuːtʃuəl] *adj.* 相互的，共有的）authentication between agents at the beginning of the session and negotiation of cryptographic（['krɪptə'ɡræfɪk] *adj.* 用密码写的）keys to be used during the session.

5　A content delivery network (CDN) is a large distributed system of servers deployed in multiple data centers in the Internet. The goal of a CDN is to serve content to end-users with high availability and high performance. CDNs serve a large fraction of the Internet content today, including Web objects (text, graphics, URLs and scripts([skrɪpts] *n.* 脚本)), downloadable objects (media files, software, documents), applications (e-commerce, portals（['pɔːtlz] *n.* 门户）), live streaming media, on-demand streaming media, and social networks.

environment outside of a wiring closet.
- Rack mounted —a switch that mounts in an equipment rack.
- Chassis—with swappable module cards.
- DIN rail mounted—normally seen in industrial environments or panels.

4.2 Configuration Options

Unmanaged switches — these switches have no configuration interface or options. They are plug and play[1]. They are typically the least expensive switches found in home, SOHO, or small businesses. They can be desktop or rack mounted.

Managed switches — these switches have one or more methods to modify the operation of the switch. Common management methods include: a command-line interface (CLI) accessed via serial console, telnet or secure shell, an embedded simple network management protocol (SNMP) agent allowing management from a remote console or management station, or a Web interface for management from a Web browser. Examples of configuration changes that one can do from a managed switch include: enabling features such as spanning tree protocol, setting port bandwidth, creating or modifying virtual LANs (VLANs), etc. Two sub-classes of managed switches are marketed today:

- Smart (or intelligent) switches — these are managed switches with a limited set of management features. Likewise "Web-managed" switches are switches which fall in a market niche between unmanaged and managed. For a price much lower than a fully managed switch they provide a web interface (and usually no CLI access) and allow configuration of basic settings, such as VLANs, port-bandwidth and duplex.
- Enterprise managed (or fully managed) switches — these have a full set of management features, including CLI, SNMP agent, and Web interface. They may have additional features to manipulate configurations, such as the ability to display, modify, backup and restore configurations. Compared with smart switches, enterprise switches have more features that can be customized or optimized, and are generally more expensive than smart switches. Enterprise switches are typically found in networks with larger number of switches and connections, where centralized management is a significant savings in administrative time and effort. A stackable switch is a version of enterprise-managed switch.

Traffic monitoring on a switched network— unless port mirroring or other methods such as RMON[2], SMON or sFlow[3] are implemented in a switch, it is difficult to monitor traffic that is

---

1 In computing, a plug and play device or computer bus, is one with a specification that facilitates the discovery of a hardware component in a system without the need for physical device configuration or user <u>intervention</u> ([ˌɪntə(ː)'venʃən] *n.* 干涉) in resolving resource conflicts.

2 The remote network monitoring (RMON) was developed by the IETF to support monitoring and protocol analysis of LANs.

3 sFlow is a technology for monitoring network, wireless and host devices. The sFlow.org consortium is the authoritative source for the sFlow protocol specifications: previous version of sFlow, including RFC 3176, have been <u>deprecated</u> (['deprɪkeɪt] *vt.* 不赞成，反对，轻视).

bridged using a switch because only the sending and receiving ports can see the traffic. These monitoring features are rarely present on consumer-grade switches.

Two popular methods that are specifically designed to allow a network analyst to monitor traffic are:

- Port mirroring — the switch sends a copy of network packets to a monitoring network connection.
- SMON—"switch monitoring" is described by RFC 2613 and is a protocol for controlling facilities such as port mirroring.

Another method to monitor may be to connect a layer-1 hub between the monitored device and its switch port. This will induce minor delay, but will provide multiple interfaces that can be used to monitor the individual switch port.

New Words

switch	[swɪtʃ]	n.	交换机
multi-port	['mʌltɪpɔːt]	n.	多口，多个端口
multilayer	[ˌmʌltɪ'leɪə]	n.	多层
converge	[kən'vɜːdʒ]	v.	聚合，聚集
separate	['seprət]	adj.	分开的，分离的，个别的，单独的
	['sepəreɪt]	v.	分开，隔离，分散，分别
domain	[də'meɪn]	n.	范围，区域，领域
necessitate	[nə'sesɪteɪt]	v.	成为必要
retransmission	[ˌriːtrænz'mɪʃən]	n.	转播，中继，重发
split	[splɪt]	v.	分开，分裂，分离
probability	[ˌprɒbə'bɪləti]	n.	概率，或然性，可能性
microsegmentation	['maɪkrəʊˌsegmen'teɪʃn]	n.	微段
dedicated	['dedɪkeɪtɪd]	adj.	专门的，专注的
modular	['mɒdjələ]	adj.	模块化的，组合的
analysis	[ə'næləsɪs]	n.	分析，分解
analytic	[ˌænə'lɪtɪk]	adj.	分析的，解析的
firewall	['faɪəwɔːl]	n.	防火墙
intrusion	[ɪn'truːʒn]	n.	闯入，侵扰
detection	[dɪ'tekʃn]	n.	侦查，探测
module	['mɒdjuːl]	n.	模块
combine	[kəm'baɪn]	v.	组合，（使）联合，（使）结合
analyzer	['ænəlaɪzə]	n.	分析者，分析器
forward	['fɔːwəd]	vt.	转发，转寄，运送
redundant	[rɪ'dʌndənt]	adj.	多余的，冗余的
uninterruptible	[ʌnɪntə'rʌptəbəl]	adj.	不可打断的，不可中断的

primarily	[praɪ'merəli]	adv. 首先，起初；主要地，根本上
capacity	[kə'pæsəti]	n. 容量，才能
literally	['lɪtərəli]	adv. 照字面意义，逐字地
convergence	[kən'vɜːdʒəns]	n. 集中，集合
reconverge	[ˌriːkən'vɜːdʒ]	v. 重新聚合
progression	[prə'greʃn]	n. 行进，级数
redundancy	[rɪ'dʌndənsi]	n. 冗余
buffer	['bʌfə]	n. 缓冲器
verify	['verɪfaɪ]	vt. 检验，校验
uplink	['ʌplɪŋk]	n. 向上传输，上行线，卫星上行链路
multicast	['mʌltɪkɑːst]	n. 多点传送；多播，组播
snoop	[snuːp]	vi. 探听，调查
transaction	[træn'zækʃn]	n. 办理，处理，事务
mount	[maʊnt]	v. 安装，放置
chassis	['ʃæsi]	n. 底盘
swappable	['swɒpəbl]	adj. 可替换的
panel	['pænl]	n. 面板，嵌板，仪表板
console	['kɒnsəʊl]	n. 控制台
agent	['eɪdʒənt]	n. 代理
customize	['kʌstəmaɪz]	v. 定制，用户化
optimize	['ɒptɪmaɪz]	vt. 使最优化
stackable	['stækəbl]	adj. 可堆叠的，易叠起堆放的
induce	[ɪn'djuːs]	vt. 促使，导致，引起
minor	['maɪnə]	adj. 较小的，次要的

Phrases

network switch	网络交换
network segment	网段
network bridge	网桥
switch port	交换端口
multilayer switch	多层交换
residential gateway	家庭网关
back and forth	来来往往地，来回地
full duplex	全双工
repeater hub	转发路由器
half duplex	半双工
collision probability	碰撞概率，冲突概率
in the extreme	非常，极端

commercial use	商业用途
be adequate for	适合，足够
mirror image	镜像，映像
external device	外部设备
fan out	扇出
intrusion detection system	入侵检测系统
packet sniffer	封包监听器，封包探测器
high performance	高精确性，高性能
commercial switch	商用交换机
power over Ethernet	用以太网供电
wireless access point	无线接入点
uninterruptible power supply	不间断电源（UPS）
network sensor	网络传感器
port mirroring	端口镜像，端口映射
storage area networks	存储区域网
except for…	除……以外
spanning tree protocol	生成树协议
forwarding plane	转发平面
store and forward	存储和转送
error checking	误差校验，错误校验
fall back	后退
fragment free	无分段
adaptive switching	自适应交换
within the confines of…	在……（范围）之内
network address translation	网络地址转换
load distribution	负荷分配
stateful inspection firewall	状态检测防火墙
web cache	网页快照，网页缓存
content delivery network	内容交付网络，内容分发网络
form factor	物理尺寸和形状，规格
rack mounted	安装在机架上的
equipment rack	设备架
plug and play	即插即用
management station	管理站
web browser	网络浏览器
smart switch	智能交换机
administrative time	修理准备时间，管理实施时间
stackable switch	可堆叠交换机

Abbreviations

SOHO (Small Office/Home Office) 小型办公室/家庭办公室
ATM (Asynchronous Transfer Mode) 异步传输模式
ITU (International Telecommunication Union) 国际电信联盟
STP (Shielded Twisted Pair) 屏蔽双绞线
RSTP (Rapid Spanning Tree Protocol) 快速生成树协议
MSTP (Multi-Service Transfer Platform) 多业务传送平台
SPB (Shortest Path Bridging) 最短路径桥接
VLAN (Virtual Local Area Network) 虚拟局域网
IGMP (Internet Group Management Protocol) 因特网组管理协议
TCP (Transfer Control Protocol) 传输控制协议
URL (Uniform Resource Locator) 统一资源定位符
CLI (Command-Line Interface) 命令行界面
SNMP (Simple Network Management Protocol) 简单网络管理协议
RMON (Remote Network MONitoring) 远程网络监控
SMON (Switch Monitoring) 交换机监控
RFC (Request For Comments) 请求评议，请求注解

Exercises

【EX.6】 **Answer the following questions according to the text.**

1. What is a network switch?
2. Which is more intelligent, a switch or a hub? Why?
3. Where does an Ethernet switch operate?
4. What is a multilayer switch?
5. Why can network hubs be useful for fanning out data to several read-only analyzers?
6. Why can the connected device continue operating even when regular office power fails?
7. Why do packet collisions affect the entire network, limiting its capacity?
8. What are the four forwarding methods a bridge can use?
9. What are the two sub-classes of managed switches marketed today?
10. What are the two popular methods that are specifically designed to allow a network analyst to monitor traffic?

【EX.7】 **Translate the following terms or phrases from English into Chinese and vice versa.**

1.	network segment	1.	
2.	multilayer switch	2.	
3.	network bridge	3.	
4.	full duplex	4.	
5.	half duplex	5.	

6.	fan out	6.	
7.	wireless access point	7.	
8.	port mirroring	8.	
9.	error checking	9.	
10.	adaptive switching	10.	
11.	即插即用	11.	
12.	智能交换机	12.	
13.	*adj.* 模块化的，组合的	13.	
14.	*n.* 闯入，侵扰	14.	
15.	*n.* 防火墙	15.	

【EX.8】Translate the following sentences into Chinese.

1. The viability of multilayer switches depends on the protocol supported.
2. The majority of these cable networks have been upgraded to broadband.
3. It separates the user's name from the domain name.
4. By default, this means that automatic retransmission is enabled.
5. The platform is built to a modular design, with good reusability.
6. A firewall acts like a virtual security guard for your network.
7. Memory organization relates to internal memory capacity and structure.
8. What is the initial capacity of the following string buffer?
9. Some UNIX system operations must be performed at the console.
10. Generally, wireless LAN is constituted by access point and workstation with wireless network card.

Reading Materials

Router

A router is a device that forwards data packets between computer networks, creating an overlay internetwork[1]. A router is connected to two or more data lines from different networks. When a data packet comes in one of the lines, the router reads the address information in the packet to determine its ultimate destination. Then, using information in its routing table[2] or routing policy, it directs the packet to the next network on its journey. Routers perform the "traffic directing" functions on the Internet. A data packet is typically forwarded from one router to another through the networks that constitute the internetwork until it gets to its destination node.

1　internetwork[ɪnˈtɜːˈnetwɜːk] *n.* 网间网
2　routing table：路由表

The most familiar type of routers are home and small office routers that simply pass data, such as Web pages and email, between the home computers and the owner's cable or DSL modem, which connects to the Internet through an ISP. More sophisticated routers, such as enterprise routers, connect large business or ISP networks up to the powerful core routers that forward data at high speed along the optical fiber lines of the Internet backbone. Though routers are typically dedicated hardware devices, use of software-based routers has grown increasingly common.

1. Applications

When multiple routers are used in interconnected networks, the routers exchange information about destination addresses, using a dynamic routing protocol. Each router builds up a table listing the preferred routes[1] between any two systems on the interconnected networks. A router has interfaces for different physical types of network connections, (such as copper cables, fibre optic, or wireless transmission). It also contains firmware for different networking protocol standards. Each network interface uses this specialized computer software to enable data packets to be forwarded from one protocol transmission system to another.

Routers may also be used to connect two or more logical groups of computer devices known as subnets, each with a different sub-network address. The subnets addresses recorded in the router do not necessarily map directly to the physical interface connections. A router has two stages of operation called planes:

- Control plane[2]: a router records a routing table listing what route should be used to forward a data packet, and through which physical interface connection. It does this using internal preconfigured addresses, called static routes.
- Forwarding plane: the router forwards data packets between incoming and outgoing interface connections.

It routes it to the correct network type using information that the packet header contains. It uses data recorded in the routing table control plane.

Routers may provide connectivity within enterprises, between enterprises and the Internet, and between Internet service providers (ISPs) networks. The largest routers interconnect the various ISPs, or may be used in large enterprise networks. Smaller routers usually provide connectivity for typical home and office networks. Other networking solutions may be provided by a backbone wireless distribution system (WDS), which avoids the costs of introducing networking cables into buildings.

All sizes of routers may be found inside enterprises. The most powerful routers are usually found in ISPs, academic and research facilities. Large businesses may also need more powerful routers to cope with ever increasing demands of intranet data traffic. A three-layer model is in common use, not all of which need be present in smaller networks.

1　preferred rout：优先路径
2　control plane：控制面板

1.1 Access Routers

Access routers, including "small office/home office" (SOHO) models, are located at customer sites such as branch offices that do not need hierarchical routing of their own. Typically, they are optimized for low cost. Some SOHO routers are capable of running alternative free Linux-based firmware.

1.2 Distribution Routers

Distribution routers aggregate[1] traffic from multiple access routers, either at the same site, or to collect the data streams from multiple sites to a major enterprise location. Distribution routers are often responsible for enforcing quality of service across a WAN, so they may have considerable memory installed, multiple WAN interface connections, and substantial[2] onboard data processing routines. They may also provide connectivity to groups of file servers or other external networks.

1.3 Security Routers

External networks must be carefully considered as part of the overall security strategy. Separate from the router may be a firewall or VPN handling device, or the router may include these and other security functions.

1.4 Core Routers

In enterprises, a core router may provide a "collapsed backbone[3]" interconnecting the distribution tier[4] routers from multiple buildings of a campus, or large enterprise locations. They tend to be optimized for high bandwidth, but lack some of the features of edge routers[5].

1.5 Internet Connectivity and Internal Use

Routers intended for ISP and major enterprise connectivity usually exchange routing information using the border gateway protocol (BGP). RFC 4098 standard defines the types of BGP-protocol routers according to the routers' functions:

- Edge router: also called a provider edge router, is placed at the edge of an ISP network. The router uses external BGP to EBGP[6] protocol routers in other ISPs, or a large enterprise autonomous system.
- Subscriber edge router: also called a customer edge router, is located at the edge of the subscriber's network, it also uses EBGP to its provider's Autonomous System. It is typically used in an (enterprise) organization.
- Inter-provider border router: interconnecting ISPs, is a BGP-protocol router that maintains BGP sessions with other BGP protocol routers in ISP autonomous systems[7].

---

1. aggregate ['ægrɪgeɪt] v. 聚集，集合
2. substantial [səb'stænʃəl] adj. 坚固的，实质的
3. collapsed backbone: 折叠式骨干
4. tier [tɪə] n. 行，排，层
5. edge router：边缘路由器
6. EBGP（External Border Gateway Protocol）：外部边界网关协议
7. autonomous system：自治系统

- Core router: a core router resides within an autonomous system as a backbone to carry traffic between edge routers.
- Within an ISP: in the ISPs autonomous system, a router uses internal BGP protocol to communicate with other ISP edge routers, other intranet core routers, or the ISPs intranet provider border routers.
- "Internet backbone": the Internet no longer has a clearly identifiable backbone, unlike its predecessor networks. The major ISPs system routers make up what could be considered to be the current Internet backbone core. ISPs operate all four types of the BGP routers described here. An ISP "core" router is used to interconnect its edge and border routers. Core routers may also have specialized functions in virtual private networks based on a combination of BGP and multi-protocol label switching[1] protocols.
- Port forwarding: routers are also used for port forwarding between private internet connected servers.
- Voice/data/fax/video processing routers: commonly referred to as access servers or gateways, these devices are used to route and process voice, data, video, and fax traffic on the internet.

2. Forwarding

For pure internet protocol (IP) forwarding function, a router is designed to minimize the state information associated with individual packets. The main purpose of a router is to connect multiple networks and forward packets destined either for its own networks or other networks. A router is considered a layer 3 device because its primary forwarding decision is based on the information in the layer 3 IP packet, specifically the destination IP address. This process is known as routing. When each router receives a packet, it searches its routing table to find the best match between the destination IP address of the packet and one of the network addresses in the routing table. Once a match is found, the packet is encapsulated in the layer 2 data link frame for that outgoing interface. A router does not look into the actual data contents that the packet carries, but only at the layer 3 addresses to make a forwarding decision, plus optionally other information in the header for hints on. Once a packet is forwarded, the router does not retain any historical information about the packet, but the forwarding action can be collected into the statistical data, if so configured.

Forwarding decisions can involve decisions at layers other than layer 3. A function that forwards based on layer 2 information is properly called a bridge. This function is referred to as layer 2 bridging, as the addresses it uses to forward the traffic are layer 2 addresses (e.g. MAC addresses on Ethernet).

Besides making decisions as which interface a packet is forwarded to, which is handled primarily via the routing table, a router also has to manage congestion, when packets arrive at a

---

1　multi-protocol label switching：多协议标签交换

rate higher than the router can process. Three policies commonly used in the Internet are tail drop[1], random early detection (RED[2]), and weighted random early detection (WRED[3]). Tail drop is the simplest and most easily implemented; the router simply drops packets once the length of the queue[4] exceeds the size of the buffers in the router. RED monitors the average queue size and drops packets based on statistical probabilities[5]. If the buffer is almost empty, all incoming packets are accepted. As the queue grows, the probability for dropping an incoming packet grows too. When the buffer is full, the probability has reached 1 and all incoming packets are dropped. WRED requires a weight on the average queue size to act upon when the traffic is about to exceed the preconfigured size, so that short bursts[6] will not trigger random drops.

Another function a router performs is to decide which packet should be processed first when multiple queues exist. This is managed through quality of service (QoS[7]), which is critical when voice over IP is deployed, so that delays between packets do not exceed 150 ms to maintain the quality of voice conversations.

Yet another function a router performs is called policy-based routing[8] where special rules are constructed to override the rules derived from the routing table when a packet forwarding decision is made.

These functions may be performed through the same internal paths that the packets travel inside the router. Some of the functions may be performed through an application-specific integrated circuit (ASIC[9]) to avoid overhead caused by multiple CPU cycles, and others may have to be performed through the CPU as these packets need special attention that cannot be handled by an ASIC.

参考译文

Text A 组网硬件

组网硬件包括全部的计算机、外部设备、接口卡及进行数据处理和网内通信所需的其他设备（见图 5-1）。

（图略）

本节提供以下设备的相关信息：

- 文件服务器；

---

1 tail drop：尾部丢弃
2 random early detection (RED)：随机早期检测
3 weighted random early detection (WRED)：加权随机早期检测
4 queue [kju:] *n.* 队列
5 statistical probability：统计概率
6 short burst：短脉冲群
7 quality of service (QoS)：服务质量
8 policy-based routing：策略路由
9 application-specific integrated circuit (ASIC)：专用集成电路

- 工作站；
- 网卡；
- 集中器/集线器；
- 中继器；
- 网桥；
- 路由器。

1. 文件服务器

文件服务器是大多数网络的核心。它是一台速度很快的计算机，带有大容量随机存储器和存储空间并配有快速网卡。该计算机装载了网络操作系统及应用软件，还装载了需要共享的数据文件。

文件服务器管理网络各节点之间的信息通信。例如，它可以同时把字处理软件发送给一个工作站、从另一个工作站接收数据库文件并存储一个电子邮件。这需要一台能够存储大量的信息并快速地分享这些信息的服务器。服务器至少要满足以下要求：

- 75MHz 或以上的微处理器；
- 容量至少有 4GB 的快速硬盘；
- 一个 RAID 以便在磁盘损坏后保存数据；
- 备份磁带机；
- 多个扩充插槽；
- 快速网卡；
- 至少 32 MB 的随机存储器。

2. 工作站

网络中所有连接到文件服务器的计算机都叫作工作站。一个典型的工作站是配有网卡、网络软件和适当电缆的计算机。工作站并不必须配置软盘驱动器或硬盘驱动器，因为文件可以存放在文件服务器上。几乎任何计算机都可以作为网络工作站。

3. 网卡

网卡提供网络和工作站之间的物理连接。大多数网卡都是内置的，插在计算机内的扩展槽上。有些计算机，如 Mac 这一类的计算机，使用连接到串口或 SCSI 接口的扩展箱。笔记本计算机通常使用连接到并口的外部 LAN 适配器或者插到 PCMCIA 槽的网卡。

网卡是决定网络速度和性能的重要因素。为你所用的工作站配备最快的网卡是个好主意。

三种最常见的网络接口连接是以太网卡、LocalTalk 连接器和令牌网卡。一家国际数据公司指出：以太网卡最流行，其次是令牌网卡和 LocalTalk 连接器。

3.1 以太网卡

虽然许多计算机（如麦金塔电脑）可以选择预装以太网卡，但通常网卡与计算机是分别购买的。以太网卡要么连接同轴电缆，要么连接双绞线（或者两者都行）。如果要连接同轴电缆，就用 BNC。如果要连接双绞线，就用 RJ-45。有些网卡带有 AUI 连接器。这可以用来把同轴电缆、双绞线或光缆连接到以太网卡。当使用这种模式时，就需要一个连接到工作站上的外部收发器。
3.2 令牌网卡
令牌网卡看上去类似于以太网卡。明显的不同在于该卡背面的连接器类型。通常令牌网卡有九针 DIN 型连接器把该卡与网络电缆相连。

4. 集中器/集线器
集中器是一个为工作站、服务器和外部设备提供中心连接点的设备。在星状拓扑中，使用双绞线把每个工作站连接到中心集中器上。集线器是内置了多口卡的多槽集中器，当网络规模增加时，可以提供新增的访问。有些连接器是被动式的。它们让信息无改变地从一个计算机传输到另一个计算机。大多数连接器是主动式的，当把信号从一个设备传输到另一个设备时，增强信号。主动式连接器（如中继器）用来扩展网络的长度。集中器：
- 通常带有 8 个、12 个或 24 个 RJ-45 端口；
- 通常用于星状或星状-环状拓扑；
- 有专门的端口管理软件销售；
- 也叫集线器；
- 通常安装在标准化的金属架上，该架上也安装网络调制解调器、网桥或路由器。

5. 中继器
当信号在电缆中传输时，强度会降低。中继器是一个增强所通过网络信号的设备。它增强所接收的信号并转发它。中继器可以是一个独立设备，也可以集成到集中器中。当网络电缆的总长度超过所用标准规定的长度时，就可以使用中继器。

一个好例子就是在星状拓扑的非屏蔽双绞线局域网中使用中继器。非屏蔽双绞线电缆的长度限定在 100 米内。最常用的配置是用双绞线把每个工作站连接到多端口主动型集中器上。集中器重新生成所有经过它的信号，使网络电缆的总长度超过 100 米的限制。

6. 网桥
网桥是一个设备，可以把一个大网段分割成两个更小、更有效的网段。如果要给旧的布线方案中增加新网络并使该新网络最先进，那么网桥可以将新旧网络相连。

网桥监控网络两边通过的信息，以便把信息包传到正确的位置。大多数网桥可以"监听"网络并自动计算出网桥两边每个计算机的地址。如果必要，网桥可以检查每个信息，并把它广播给网络的另一端。

网桥管理流量以便使网络两边的性能最佳。也可以说网桥就像高峰期十字路口的交通警察。它管理网络两边的信息流，阻止不必要的信息通过。网桥可以连接不同类型的电缆或物理拓扑，但是它们必须用在使用相同协议的网络之中。

7. 路由器
路由器把信息从一个网络传输给另一个网络，它类似于超级智能的网桥。路由器根据目的地址和起始地址选择最佳路径来传递消息。路由器可以引导流量以防止正面冲突。它也有足够的智能知道何时把流量引导到僻径和捷径上。

尽管网桥知道网络每一边全部计算机的地址，但路由器知道网络上计算机、网桥和其他路由器的地址。路由器甚至可以"监听"整个网络以便确定哪一段是最繁忙的——这样它们可以引导数据绕过这些路段，直到不再繁忙为止。

如果有连接到因特网的学校局域网，就需要买一个路由器。在这种情况下，路由器就在局域网和因特网之间翻译信息。它也可以确定通过因特网之间发送数据的最佳路径。路由器可以：

- 有效引导信号流量；
- 在两个协议之间发送消息；
- 在线性总线、星状总线和星状-环状拓扑之间发送消息；
- 在光缆、同轴电缆和双绞线之间传输消息。

Unit 6

Text A

Wireless Sensor Network

A wireless sensor network (WSN) consists of spatially distributed autonomous sensors to monitor physical or environmental conditions, such as temperature, sound, pressure, etc. and to cooperatively pass their data through the network to a main location. The more modern networks are bidirectional, and they are also enabling control of sensor activity. The development of wireless sensor networks was motivated by military applications such as battlefield surveillance; today such networks are used in many industrial and consumer applications, such as industrial process monitoring and control, machine health monitoring, and so on.

The WSN is built of "nodes"—from a few to several hundreds or even thousands, where each node is connected to one (or sometimes several) sensors. Each such sensor network node has typically several parts: a radio transceiver with an internal antenna or connection to an external antenna, a microcontroller, an electronic circuit for interfacing with the sensors and an energy source, usually a battery or an embedded form of energy harvesting. A sensor node might vary in size from that of a shoebox down to the size of a grain of dust, although functioning "motes" of genuine microscopic dimensions have yet to be created. The cost of sensor nodes is similarly variable, ranging from a few to hundreds of dollars, depending on the complexity of the individual sensor nodes. Size and cost constraints on sensor nodes result in corresponding constraints on resources such as energy, memory, computational speed and communications bandwidth. The topology of the WSNs can vary from a simple star network to an advanced multihop wireless mesh network. The propagation technique between the hops of the network can be routing or flooding.

In computer science and telecommunications, wireless sensor networks are an active research area with numerous workshops and conferences arranged each year.

1. Characteristics

The main characteristics of a WSN include:
- Power consumption constrains nodes using batteries or energy harvesting.
- Ability to cope with node failures.
- Mobility of nodes.
- Communication failures.
- Heterogeneity of nodes.
- Scalability to large scale of deployment.
- Ability to withstand harsh environmental conditions.
- Ease of use.
- Power consumption.

Sensor nodes can be imagined as small computers, extremely basic in terms of their interfaces and their components. They usually consist of a processing unit with limited computational power and limited memory, sensors or MEMS (including specific conditioning circuitry), a communication device (usually radio transceivers or alternatively optical), and a power source usually in the form of a battery. Other possible inclusions are energy harvesting modules, secondary ASICs, and possibly secondary communication devices (e.g. RS-232 or USB).

The base stations are one or more components of the WSN with much more computational, energy and communication resources. They act as a gateway between sensor nodes and the end user as they typically forward data from the WSN on to a server. Other special components in routing based networks are routers, which are designed to compute, calculate and distribute the routing tables.

2. Platforms

2.1 Standards and Specifications

Several standards are currently either ratified or under development by organizations including WAVE2M[1] for wireless sensor networks. There are a number of standardization bodies in the field of WSNs. The IEEE focuses on the physical and MAC layers; the Internet Engineering Task Force works on layers 3 and above. In addition to these, bodies such as the International Society of Automation provide vertical solutions, covering all protocol layers. Finally, there are also several non-standard, proprietary mechanisms and specifications.

Standards are used far less in WSNs than in other computing systems, which makes most systems incapable of direct communication between different systems. However, predominant standards commonly used in WSN communications include:

---

1 WAVE2M is an international nonprofit （[ˌnɒnˈprɒfɪt] *adj*. 非营利的）standard development organization founded to promote the global use and enhancement of WAVE2M, an emerging wireless communication technology standard for ultra-low-power （超低功耗）and long-range devices.

- WirelessHART[1]
- IEEE 1451[2]
- ZigBee / 802.15.4
- ZigBee IP
- 6LoWPAN[3]

2.2 Hardware

One major challenge in a WSN is to produce low cost and tiny sensor nodes. There are an increasing number of small companies producing WSN hardware and the commercial situation can be compared to home computing in the 1970s. Many of the nodes are still in the research and development stage, particularly their software. Also inherent to sensor network adoption is the use very low power methods for data acquisition.

2.3 Software

Energy is the scarcest resource of WSN nodes, and it determines the lifetime of WSNs. WSNs are meant to be deployed in large numbers in various environments, including remote and wild regions, where ad-hoc communications are a key component. For this reason, algorithms and protocols need to address the following issues: lifetime maximization, robustness and fault tolerance and self-configuration.

Energy/power consumption of the sensing device should be minimized and sensor nodes should be energy efficient since their limited energy resource determines their lifetime. To conserve power the node should shut off the radio power supply when not in use.

Some of the important topics in WSN software research are operating systems, security and mobility.

Operating systems for wireless sensor network nodes are typically less complex than general-purpose operating systems. They more strongly resemble embedded systems for two reasons. First, wireless sensor networks are typically deployed with a particular application in mind rather than as a general platform. Second, a need for low costs and low power leads most wireless sensor nodes to have low-power microcontrollers ensuring that mechanisms, such as virtual memory, are either unnecessary or too expensive to implement.

It is, therefore, possible to use embedded operating systems such as eCos or μC/OS for sensor networks. However, such operating systems are often designed with real-time properties.

---

1 WirelessHART is a wireless sensor networking technology based on the highway addressable remote transducer protocol (HART, 可寻址远程传感器高速通道).

2 IEEE 1451 is a set of smart transducer interface standards developed by the Institute of Electrical and Electronics Engineers (IEEE) Instrumentation and Measurement Society's Sensor Technology Technical Committee that describe a set of open, common, network-independent communication interfaces for connecting transducers (sensors or actuators) to microprocessors, instrumentation ([ˌɪnstrumenˈteɪʃən] n. 仪器, 仪表) systems, and control/field networks.

3 6LoWPAN is an acronym of IPv6 over low power wireless personal area networks（用 IPv6 的低功率无线个人局域网络）. 6LoWPAN is the name of a working group in the Internet area of the IETF. The 6LoWPAN concept originated from the idea that "the internet protocol could and should be applied even to the smallest devices" and that low-power devices with limited processing capabilities should be able to participate in the Internet of Things.

TinyOS is perhaps the first operating system specifically designed for wireless sensor networks. TinyOS is based on an event-driven programming[1] model instead of multithreading[2]. TinyOS programs are composed of event handlers and tasks with run-to-completion semantics. When an external event occurs, such as an incoming data packet or a sensor reading, TinyOS signals the appropriate event handler to handle the event. Event handlers can post tasks that are scheduled by the TinyOS kernel some time later.

LiteOS is a newly developed OS for wireless sensor networks, which provides UNIX-like abstraction and support for the C programming language.

Contiki is an OS which uses a simpler programming style in C while providing advances such as 6LoWPAN and Protothreads[3].

3. Other Concepts

3.1 Distributed Sensor Network

If a centralized architecture is used in a sensor network and the central node fails, the entire network will collapse.However the reliability of the sensor network can be increased by using distributed control architecture.

Distributed control is used in WSNs for the following reasons:
- Sensor nodes are prone to failure.
- For better collection of data.
- To provide nodes with backup in case of failure of the central node.
- There is also no centralized body to allocate the resources and they have to be self organized.

3.2 Data Integration and Sensor Web

The data gathered from wireless sensor networks is usually saved in the form of numerical data in a central base station. Additionally, the Open Geospatial Consortium (OGC)[4] is specifying standards for interoperability interfaces and metadata encodings that enable real time integration of heterogeneous sensor Webs into the Internet, allowing any individual to monitor or control wireless sensor networks through a Web browser.

---

1 In computer programming, event-driven programming (EDP) or event-based programming（基于事件编程）is a programming paradigm（['pærədaɪm] *n.* 范例）in which the flow of the program is determined by events—e.g., sensor outputs or user actions (mouse clicks, key presses) or messages from other programs or threads（[θredz] *n.* 线程）.

2 In computer science, a thread of execution is the smallest sequence of programmed instructions（程序指令）that can be managed independently by an operating system scheduler. A thread is a light-weight process.

3 A protothread is a low-overhead mechanism for concurrent（[kən'kʌrənt] *adj.* 并发的，协作的）programming.Protothreads function as stackless, lightweight threads providing a blocking context cheaply using minimal memory per protothread (on the order of single bytes).

4 The Open Geospatial Consortium (OGC), an international voluntary（['vɒləntri] *adj.* 自愿的）consensus standards organization, originated in 1994. In the OGC, more than 400 commercial, governmental, nonprofit and research organizations worldwide collaborate in a consensus process encouraging development and implementation of open standards for geospatial content and services, GIS data processing and data sharing.

3.3 In-network Processing

To reduce communication costs some algorithms remove or reduce nodes redundant sensor information and avoid forwarding data that is of no use. As nodes can inspect the data they forward they can measure the average or directionality of readings from other nodes. For example, in sensing and monitoring applications, it is generally the case that neighbouring sensor nodes monitoring an environmental feature typically register similar values. This kind of data redundancy due to the spatial correlation between sensor observations inspires the techniques for in-network data aggregation and mining.

New Words

sensor	['sensə]	n.	传感器
spatially	['speɪʃəli]	adv.	空间地
autonomous	[ɔː'tɒnəməs]	adj.	自治的
temperature	['temprətʃə]	n.	温度
cooperatively	[kəu'ɒpəreɪtɪvli]	adv.	合作地，协力地
bidirectional	[ˌbaɪdə'rekʃənl]	adj.	双向的
activity	[æk'tɪvəti]	n.	活动，行动
antenna	[æn'tenə]	n.	天线
battlefield	['bætlfiːld]	n.	战场，沙场
microcontroller	[ˌmaɪkrəukɒn'trəulə]	n.	微控制器
battery	['bætri]	n.	电池
mote	[məut]	n.	尘埃，微粒
genuine	['dʒenjuɪn]	adj.	真实的，真正的
microscopic	[ˌmaɪkrə'skɒpɪk]	adj.	极小的，微小的
variable	['veərɪəbl]	adj.	可变的，不定的
correspond	[ˌkɒrə'spɒnd]	vi.	符合，协调，通信，相应
computational	[ˌkɒmpju'teɪʃənl]	adj.	计算的
propagation	[ˌprɒpə'geɪʃn]	n.	（声波、电磁辐射等）传播
flooding	['flʌdɪŋ]	n.	泛洪法；涌入，流入
workshop	['wɜːkʃɒp]	n.	车间，工场
conference	['kɒnfərəns]	n.	会议，讨论会，协商会
consumption	[kən'sʌmpʃn]	n.	消费，消耗
hop	[hɒp]	v.	跳跃
heterogeneity	[ˌhetərəudʒɪ'niːɪti]	n.	异种，异质，不同成分
scalability	[skeɪlə'bɪlɪti]	n.	可测量性
withstand	[wɪð'stænd]	vt.	抵挡，经受住
harsh	[hɑːʃ]	adj.	苛刻的；荒芜的
optical	['ɒptɪkl]	adj.	视力的，光学的

platform	['plætfɔːm]	*n.* 平台
ratify	['rætɪfaɪ]	*vt.* 批准，认可
proprietary	[prə'praɪətri]	*adj.* 所有的
		n. 所有者，所有权
incapable	[ɪn'keɪpəbl]	*adj.* 无能力的，不能的
predominant	[prɪ'dɒmɪnənt]	*adj.* 支配的，主要的，有影响的
tiny	['taɪni]	*adj.* 很少的，微小的
scarce	[skeəs]	*adj.* 缺乏的，不足的，稀有的
ad-hoc	[ˌæd'hɒk]	*adj.* 特别
maximization	[ˌmæksɪmaɪ'zeɪʃn]	*n.* 最大值化，极大值化
robustness	[rəʊ'bʌstnəs]	*n.* 坚固性，健壮性，鲁棒性
conserve	[kən'sɜːv]	*vt.* 保存，保藏
particular	[pə'tɪkjələ]	*n.* 细节，详细
		adj. 特殊的，特别的，独特的
property	['prɒpəti]	*n.* 性质，特性；财产，所有权
resemble	[rɪ'zembl]	*vt.* 像，类似
multithreading	[ˌmʌltɪ'θrɛdɪŋ]	*n.* 多线程
task	[tɑːsk]	*n.* 任务，作业
		v. 分派任务
handler	['hændlə]	*n.* 处理者，处理器
allocate	['æləkeɪt]	*vt.* 分派，分配
protothread	['prəʊtəʊθrɛd]	*n.* 轻量级线程
metadata	['metədeɪtə]	*n.* 元数据
heterogeneous	[ˌhetərə'dʒiːniəs]	*adj.* 不同种类的，异类的
neighbouring	['neɪbərɪŋ]	*adj.* 附近的，毗邻的
spatial	['speɪʃl]	*adj.* 空间的
correlation	[ˌkɒrə'leɪʃn]	*n.* 相互关系，相关性
inspire	[ɪn'spaɪə]	*vt.* 激发，产生

Phrases

battlefield surveillance	战场侦察，战场监视
machine health monitoring	机器的健康监测
electronic circuit	电子电路
energy harvesting	能量采集
sensor node	传感器节点
a grain of	一粒；一点点，一些
communication bandwidth	通信带宽
star network	星状网络

multihop wireless mesh network	多跳无线网状网络
power consumption	能量消耗，功率消耗，动力消耗
cope with	应付
environmental condition	环境条件，环境状况
processing unit	处理部件，处理器
base station	基站，基地
special component	专有部件，专用附件
focus on	致力于；使聚焦于；对（某事或做某事）予以注意；把……作为兴趣中心
Internet Engineering Task Force	因特网工程工作小组
International Society of Automation	国际自动化学会
research and development	研究与开发，研发
data acquisition	数据获取
fault tolerance	容错
sensing device	灵敏元件，传感器
power supply	电源
embedded system	嵌入式系统
virtual memory	虚拟内存
embedded operating system	嵌入式操作系统
event-driven programming model	事件驱动编程模型
be composed of	由……组成
event handler	事件处理器
be prone to…	有……的倾向，易于
in case of	假设，万一
in the form of…	以……的形式

Abbreviations

WSN (Wireless Sensor Network)	无线传感器网络
MEMS (Microelectromechanical System)	微机电系统
ASIC (Application Specific Integrated Circuit)	专用集成电路
USB (Universal Serial Bus)	通用串行总线
OGC (Open Geospatial Consortium)	开放式地理信息系统协会

Analysis of Difficult Sentences

[1] A sensor node might vary in size from that of a shoebox down to the size of a grain of dust, although functioning "motes" of genuine microscopic dimensions have yet to be created.

本句中，that 指代 the size。although functioning "motes" of genuine microscopic dimensions have yet to be created 是一个让步状语从句，修饰谓语 might vary。

[2] Other special components in routing based networks are routers, which are designed to

compute, calculate and distribute the routing tables.

本句中，which are designed to compute, calculate and distribute the routing tables 是一个非限定性定语从句，对 routers 进行补充说明。

[3] Standards are used far less in WSNs than in other computing systems, which makes most systems incapable of direct communication between different systems.

本句中，which makes most systems incapable of direct communication between different systems 是一个非限定性定语从句，对它前面的句子进行补充说明。在该从句中，incapable of direct communication between different systems 是一个形容词短语，作宾语 most systems 的补足语。

[4] WSNs are meant to be deployed in large numbers in various environments, including remote and wild regions, where ad-hoc communications are a key component.

本句中，where ad-hoc communications are a key component 是一个非限定性定语从句，对 remote and wild regions 进行补充说明。

[5] Additionally, the Open Geospatial Consortium (OGC) is specifying standards for interoperability interfaces and metadata encodings that enable real time integration of heterogeneous sensor Webs into the Internet, allowing any individual to monitor or control wireless sensor networks through a Web browser.

本句中, for interoperability interfaces and metadata encodings 是一个介词短语，作定语，修饰和限定 standards。that enable real time integration of heterogeneous sensor Webs into the Internet 是一个定语从句，也修饰和限定 standards。allowing any individual to monitor or control wireless sensor networks through a Web browser 是一个现在分词短语，作结果状语。

Exercises

【EX.1】 Answer the following questions according to the text.

1. What does a wireless sensor network (WSN) consist of?
2. What can sensor nodes be imagined as?
3. What do base stations act as?
4. What do predominant standards commonly used in WSN communications include?
5. What is one major challenge in a WSN?
6. What is the scarcest resource of WSN nodes?
7. What are some of the important topics in WSN software research?
8. What is perhaps the first operating system specifically designed for wireless sensor networks? What is it based on?
9. What are the reasons for the use of distributed control in WSNs?
10. How is the data gathered from wireless sensor networks usually saved?

【EX.2】 Translate the following terms or phrases from English into Chinese and vice versa.

1. sensor node   1. _____
2. communication bandwidth   2. _____

3.	multihop wireless mesh network	3.	
4.	star network	4.	
5.	fault tolerance	5.	
6.	base station	6.	
7.	sensing device	7.	
8.	embedded operating system	8.	
9.	microcontroller	9.	
10.	computational	10.	
11.	n. 平台	11.	
12.	n. 坚固性，健壮性，鲁棒性	12.	
13.	n. 多线程	13.	
14.	n. 元数据	14.	
15.	n. 处理者，处理器	15.	

【EX.3】 Translate the following sentences into Chinese.

1. Border gateway protocol (BGP) provides loop-free inter domain routing between autonomous systems.
2. In many data communications ways, cable is an ideal bidirectional Internet access technology.
3. Computational grid is a burgeoning high performance computing technique.
4. The new algorithm is also applicable to general relational database.
5. Robustness analysis attracts more and more attention in these years.
6. An event handler is the code you write to respond to event.
7. Metadata is the soul of data warehouse.
8. The core idea of heterogeneous database inter-operation is data sharing and transparent accessing.
9. The standards of Web service solved interoperability puzzles among heterogeneous information systems.
10. It is essential to consider system level power consumption when designing current mobile devices.

【EX.4】 Complete the following passage with appropriate words in the box.

touching	voltmeter	aware	liquid	measure
indicates	sensitivities	manufactured	responds	converts

A sensor (also called detector) is a converter that measures a physical quantity and converts it into a signal which can be read by an observer or by an (today mostly electronic) instrument.

For example, a mercury-in-glass thermometer ___1___ the measured temperature into expansion and contraction of a ___2___ which can be read on a calibrated glass tube. A thermocouple converts temperature to an output voltage which can be read by a ___3___. For accuracy, most sensors are calibrated against known standards.

Sensors are used in everyday objects such as touch-sensitive elevator buttons (tactile sensor) and lamps which dim or brighten by ___4___ the base. There are also innumerable applications for sensors of which most people are never ___5___. Applications include cars, machines, aerospace, medicine, manufacturing and robotics.

A sensor is a device which receives and ___6___ to a signal when touched. A sensor's sensitivity ___7___ how much the sensor's output changes when the measured quantity changes. For instance, if the mercury in a thermometer moves 1 cm when the temperature changes by 1 ℃, the sensitivity is 1cm/°C. Sensors that ___8___ very small changes must have very high ___9___. Sensors also have an impact on what they measure; for instance, a room temperature thermometer inserted into a hot cup of liquid cools the liquid while the liquid heats the thermometer. Sensors need to be designed to have a small effect on what is measured; making the sensor smaller often improves this and may introduce other advantages. Technological progress allows more and more sensors to be ___10___ on a microscopic scale as microsensors using MEMS technology. In most cases, a microsensor reaches a significantly higher speed and sensitivity compared with macroscopic approaches.

【EX.5】 Translate the following passage into Chinese.

Sensor Network

A sensor network is a group of specialized transducers with a communications infrastructure intended to monitor and record conditions at diverse locations. Commonly monitored parameters are temperature, humidity, pressure, wind direction and speed, illumination intensity, vibration intensity, sound intensity, power-line voltage, chemical concentrations, pollutant levels and vital body functions.

A sensor network consists of multiple detection stations called sensor nodes, each of which is small, lightweight and portable. Every sensor node is equipped with a transducer, microcomputer, transceiver and power source. The transducer generates electrical signals based on sensed physical effects and phenomena. The microcomputer processes and stores the sensor output. The transceiver, which can be hard-wired or wireless, receives commands from a central computer and transmits data to that computer. The power for each sensor node is derived from the electric utility or from a battery.

Potential applications of sensor networks include:
- Industrial automation.
- Automated and smart homes.
- Video surveillance.
- Traffic monitoring.

- Medical device monitoring.
- Monitoring of weather conditions.
- Air traffic control.
- Robot control.

Text B

WiFi

1. What Is WiFi?

WiFi is a wireless networking technology that allows devices such as computers (laptops and desktops), mobile devices (smart phones and wearables), and other equipment (printers and video cameras) to interface with the Internet. It allows these devices—and many more—to exchange information with one another, creating a network.

Internet connectivity occurs through a wireless router[1]. When you access WiFi, you are connecting to a wireless router that allows your WiFi-compatible devices to interface with the Internet.

2. WiFi Standards

On the technical side, the IEEE 802.11 standard defines the protocols that enable communications with current WiFi-enabled wireless devices, including wireless routers and wireless access points. Each standard is an amendment that was ratified over time. The standards operate on varying frequencies, deliver different bandwidth, and support different numbers of channels.

WiFi uses 802.11 networking standards, which come in several flavors and have evolved over the decades:

- 802.11b (introduced in 1999) is the slowest and least expensive standard. For a while, its cost made it popular, but now it's less common as faster standards become less expensive.
- 802.11a (introduced after 802.11b) transmits at 5 GHz and can move up to 54 megabits of data per second. It uses orthogonal frequency-division multiplexing(OFDM), a more efficient coding technique that splits that radio signal into several sub-signals before they reach a receiver. This greatly reduces interference.
- 802.11g transmits at 2.4 GHz like 802.11b, but it's a lot faster. It can handle up to 54 megabits of data per second. It is faster because it uses the same OFDM coding as 802.11a.

---

1 A wireless router is a device that performs the functions of a router and also includes the functions of a wireless access point. It is used to provide access to the Internet or a private computer network. Depending on the manufacturer and model, it can function in a wired local area network, in a wireless-only LAN, or in a mixed wired and wireless network.

- 802.11n (introduced in 2009) is backward compatible with a, b and g. It significantly improved speed and range over its predecessors. For instance, although 802.11g theoretically moves 54 megabits of data per second, it only achieves real-world speeds of about 24 megabits of data per second because of network congestion. 802.11n, however, reportedly can achieve speeds as high as 140 megabits per second. 802.11n can transmit up to four streams of data, each at a maximum of 150 megabits per second, but most routers only allow for two or three streams.
- 802.11ac came on the scene around 2014, and operates exclusively at a 5 GHz frequency. 802.11ac is backward compatible with 802.11n (and therefore the others, too), with n on the 2.4 GHz band and ac on the 5 GHz band. It is less prone to interference and far faster than its predecessors, pushing a maximum of 450 megabits per second on a single stream, although real-world speeds may be lower. Like 802.11n, it allows for transmission on multiple spatial streams — up to eight, optionally. It is sometimes called 5G because of its frequency band, sometimes gigabit WiFi because of its potential to exceed a gigabit per second on multiple streams and sometimes very high throughput (VHT) for the same reason.
- 802.11ax, also known as WiFi 6, came to the industry in 2019. This standard extends the capabilities of 802.11ac in a few key ways. First of all, the new routers allow an even higher data flow rate, up to 9.2 Gbps (gigabits per second). WiFi 6 also allows manufacturers to install many more antennas on one router, accepting multiple connections at once without any worry of interference and slowdown. Some new devices also connect on a higher 6 GHz band, which is about 20 percent faster than 5 GHz in ideal conditions.
- 802.11be (or WiFi 7) is projected to be the standard by 2024, and should offer even better range, more connections and faster data rates than any of the previous versions.

Other 802.11 standards focus on specific applications of wireless networks, like wide area network (WAN[1]) inside vehicles or technology that allows you to move from one wireless network to another seamlessly.

WiFi radios can transmit on any frequency band. Or they can "frequency hop" rapidly between the different bands. Frequency hopping helps reduce interference and allows multiple devices to use the same wireless connection simultaneously.

As long as they all have wireless adapters, several devices can use one router to connect to the internet. This connection is convenient, virtually invisible and fairly reliable; however, if the router fails or if too many people try to use high-bandwidth applications at the same time, users can experience interference or lose their connections, although newer, faster standards like 802.11ax will help with that.

---

1 Wide area network (WAN) is a connected collection of telecommunication networks distributed across a large geographic area spanning multiple cities, territories, or nations so that the component networks can exchange data within the defined WAN group.

3. WiFi Hot Spots

A WiFi hot spot is simply an area with an accessible wireless network. You can even create your own mobile hot spot using a cell phone or an external device that can connect to a cellular network. And you can always set up a WiFi network at home.

If you want to take advantage of public WiFi hot spots or your own home-based network, the first thing you'll need to do is make sure your computer has the right gear. Most new laptops and many new desktop computers come with built-in wireless transmitters, and just about all mobile devices are WiFi enabled. If your computer isn't already equipped, you can buy a wireless adapter that plugs into the PC card slot or USB port. Desktop computers can use USB adapters, or you can buy an adapter that plugs into the PCI slot inside the computer's case. Many of these adapters can use more than one 802.11 standard.

Once you've installed a wireless adapter and the drivers that allow it to operate, your computer should be able to automatically discover existing networks. This means that when you turn your computer on in a WiFi hot spot, the computer will inform you that the network exists and ask whether you want to connect to it.

4. Building a Wireless Network

If you already have several computers networked in your home, you can create a wireless network with a wireless access point. If you have several computers that are not networked, or if you want to replace your Ethernet network[1], you'll need a wireless router. This is a single unit that contains:

- a port to connect to your cable or DSL modem.
- a router.
- a firewall[2].
- a wireless access point.

A wireless router allows you to use wireless signals or Ethernet cables to connect your computers and mobile devices to one another, to a printer and to the internet. Most routers provide coverage for about 100 feet (30.5 meters) in all directions, although walls and doors can block the signal. If your home is very large, you can buy inexpensive range extenders or repeaters to increase your router's range.

Once you plug in your router, it should start working at its default settings. Most routers let you use a web interface to change your settings. You can select:

- The name of the network, known as its service set identifier (SSID). The default setting

---

1 Ethernet network refers to a type of computer network that uses Ethernet communication protocol to connect devices, such as computers, printers, and servers within a local area network (LAN). It allows the devices to communicate with each other and share resources such as files, documents, and internet access. Ethernet networks can be either wired or wireless and use various topologies such as star, bus, ring, and others to ensure smooth communication and data transfer between the devices in the network.

2 A firewall is a network security device that monitors and filters incoming and outgoing network traffic based on an organization's previously established security policies. At its most basic, a firewall is essentially the barrier that sits between a private internal network and the public Internet.

is usually the manufacturer's name.
- The channel that the router uses. Most routers use channel 6 by default. If you live in an apartment and your neighbors are also using channel 6, you may experience interference. Switching to a different channel should eliminate the problem.
- Your router's security options. Many routers use a standard, publicly available sign-on, so it's a good idea to set your own username and password.

5. How to Secure WiFi Network

Encrypt your network. Encryption scrambles the information sent through your network. That makes it harder for other people to see what you're doing or get your personal information. To encrypt your network, simply update your router settings to either WPA3 personal or WPA2 personal. WPA3 is the newer and best encryption available.

Change your router's default settings. Change the default administrative username, password, and network name to something unique. Don't use login names or passwords with your name, address, or router brand.

There are two passwords on your router that you'll need to reset:
- The WiFi network password: this is the one you use to connect your devices to the network. A unique and secure WiFi network password prevents strangers from getting onto your network.
- The router admin password: this is the one that lets you into the administrative side of the device. There, you're able to do things like change settings (including the WiFi network password). If a hacker[1] managed to log into the admin side of your router, the hacker could change the settings (including your WiFi network password). That would undo any other security steps you may be taking.

Keep your router up to date. Before you set up a new router or make updates to your existing one, visit the manufacturer's website to see if there's a newer version of the software available for download.

Turn off "remote management", WiFi protected setup (WPS), and universal plug and play (UPnP) features. Some routers have features that can be convenient but weaken your network security. For example, enabling remote access to your router's controls allows you to change settings over the internet. WPS lets you push a button on the router to connect a device to the Internet instead of entering the WiFi network password. Lastly, UPnP lets your devices find each other on the network. These features may make it easier to add devices to your network or let guests use your WiFi, but they can make your network less secure.

Set up a guest network. Many routers let you set up a guest network with a different name and password. It's a good security move for two reasons:
- Having a separate login means fewer people have your primary WiFi network password.

---

1 A hacker is a person who breaks into a computer system. The reasons for hacking can be many: installing malware, stealing or destroying data, disrupting service, and more.

- In case a guest (unknowingly) has malware on their phone or tablet, it won't get onto your primary network and your devices.

Log out as administrator. Once you've set up your router or are done changing settings, don't forget to log out as administrator. When you're logged in as administrator, you're able to change passwords and otherwise manage settings that control the security of your network. If a hacker got into your administrator account, they could easily get into your network and devices.

Turn on your router firewall. A firewall is an additional layer of protection that can help keep out viruses, malware, and even hackers. Most routers come with built-in firewalls, so check your settings to make sure your router's firewall is turned on.

New Words

laptop	['læptɒp]	n.	膝上型计算机
desktop	['desktɒp]	n.	桌面计算机
equipment	[ɪ'kwɪpmənt]	n.	设备，装备
compatible	[kəm'pætəbl]	adj.	兼容的，相容的
amendment	[ə'mendmənt]	n.	修改，修订
frequency	['fri:kwənsi]	n.	频率
megabit	['megəbɪt]	n.	兆位
sub-signal	[sʌb'sɪgnəl]	n.	子信号
backward	['bækwəd]	adj.	向后的；反向的
		adv.	向后地；相反地
predecessor	['pri:dɪsesə]	n.	前任，前辈，前身
congestion	[kən'dʒestʃən]	n.	拥挤，阻塞
optionally	['ɒpʃənəli]	adv.	随意地
gigabit	['gɪgəbɪt]	n.	吉位
slowdown	['sləʊdaʊn]	n.	减速，降速；减缓
cellular	['seljələ]	adj.	蜂窝的
public	['pʌblɪk]	adj.	公众的，公共的
slot	[slɒt]	n.	插槽
install	[ɪn'stɔːl]	vt.	安装
inform	[ɪn'fɔːm]	vt.	通知
cable	['keɪbl]	n.	电缆
switch	[swɪtʃ]	n.	转换器
		v.	转换
scramble	['skræmbl]	v.	打乱，混乱
login	['lɒgɪn]	n.	注册；登录
		v.	登录，进入系统
stranger	['streɪndʒə]	n.	陌生人

hacker	['hækə]	n. 黑客
weaken	['wiːkən]	v.（使）削弱；（使）变弱；衰减
button	['bʌtn]	n. 按钮
guest	[gest]	n. 客人，来宾
primary	['praɪmərɪ]	adj. 首要的，主要的
virus	['vaɪrəs]	n. 计算机病毒
built-in	[bɪlt ɪn]	adj. 嵌入的，内置的

Phrases

video camera	摄像机
wireless router	无线路由器
split ... into ...	把……划分为……
multiple spatial stream	多空间流
data flow rate	数据流率
gigabits per second	每秒吉位
frequency band	频带，频段
frequency hop	跳频
hot spot	热点
cellular network	蜂窝网
home-based network	家庭网络
computer's case	计算机机箱
Ethernet network	以太网
default setting	默认设置
remote access	远程访问
log out	注销
administrator account	管理员账号
turn on	打开

Abbreviations

OFDM (Orthogonal Frequency-Division Multiplexing)	正交频分复用
VHT (Very High Throughput)	超高吞吐量
PCI (Peripheral Component Interconnect)	外围部件互连
DSL (Digital Subscriber Line)	数字用户线
SSID (Service Set IDentifier)	服务集标识符
WPS (WiFi Protected Setup)	WiFi 保护设置
UPnP (Universal Plug and Play)	通用即插即用

Exercises

【EX.6】 Answer the following questions according to the text.

1. What is WiFi?
2. What does the IEEE 802.11 standard do on the technical side?
3. Why is 802.11g?
4. What is 802.11ax known as? When did it come to the industry?
5. What is a WiFi hot spot?
6. How can you create your own mobile hot spot?
7. What does a wireless router allows you to do?
8. Why is it a good idea to set your own username and password?
9. What are the two passwords on your router that you'll need to reset?
10. What are you able to do when you're logged in as administrator?

【EX.7】 Translate the following terms or phrases from English into Chinese and vice versa.

1. built-in
2. cable
3. cellular
4. compatible
5. congestion
6. equipment
7. install
8. cellular network
9. ethernet network
10. frequency band
11. 热点
12. 无线路由器
13. n. 转换器 v. 转换
14. n. 插槽
15. 远程访问

【EX.8】 Translate the following sentences into Chinese.

1. The device plugs into one of the laptop's USB ports.
2. It is significantly more compact than any comparable laptop, with no loss in functionality.
3. Data backup, user training, and performance issues must also be considered.
4. Make sure the software is fully compatible with this operating system.
5. This printer is compatible with most microcomputers.

6. Any device that's built to receive a wireless signal at a specific frequency can be overwhelmed by a stronger signal coming in on the same frequency.
7. High rate analog circuit was used to make Cu lead support thousands of megabit transmit rate.
8. The congestion phenomenon seriously affects network quality of service and resource utilization.
9. Modern mobile communication system requires high-speed and large capacity data transmission.
10. Press these two keys to switch between documents on screen.

Reading Material

IoT and the Smart City

A smart city[1] incorporates advanced technology, including a vast network of sensors and interconnected[2] devices, which gather real-time information to improve the efficiency of public services while saving money and resources.

IoT refers to the network of physical objects, such as devices, vehicles, and buildings, embedded with sensors, software, and connectivity, which enable them to collect and exchange data. These connected devices can communicate with other devices and systems, allowing them to function and share data seamlessly.

IoT forms the technical backbone of every smart city in the world, equipping them with the intelligence, interconnection, and instruments[3] needed to improve urban services, optimize resources, and reduce costs. By connecting various devices, systems, and people, IoT can provide real-time data and insights on city operations and infrastructure.

1. IoT Technologies for Smart Cities

The foundation of smart cities relies on the utilization of IoT devices and networks. These devices, in combination with[4] software solutions, user interfaces, and communication networks, enable and enhance the functioning and efficiency of smart cities.

Ultimately, the goal is to have IoT technologies interconnected, with data flowing seamlessly between devices, in order to create a truly smart city that can improve the quality of life. More specifically, these IoT technologies include radio frequency identification (RFID)[5], near field communication (NFC)[6], low-power wide-area (LPWA), wireless telecommunications, wireless sensor network (WSN), DASH7, and addressing.

1.1 Radio Frequency Identification (RFID)

Radio frequency identification (RFID) tagging devices use radio frequencies to transfer data mainly to track and identify objects and people. In a smart city, tracking the location and movement of vehicles, equipment, and even people can generate important data that can be used

---

1 smart city：智慧城市
2 interconnect [ˌɪntəkəˈnekt] v. 互相连接，互相联系
3 instrument [ˈɪnstrəmənt] n. 仪器，工具
4 in combination with：与……联合，与……结合
5 radio frequency identification (RFID)：射频识别
6 near field communication (NFC)：近场通信

to optimize the operation of transportation systems, waste collection routes, and parking availability.

Additionally, RFID technology can be used to improve security and public safety in a city by enabling real-time tracking of vehicles and individuals.

1.2 Near Field Communication (NFC)

Near field communication (NFC) is a set of standards for smartphones and other devices, like credit card readers, to establish radio communication with each other by bringing them into close proximity, typically within an inch (or a few centimeters). In smart cities, NFC is used for bidirectional short-distance communication[1], such as contactless payments[2], access control, and electronic ticketing.

1.3 Low Power Wide Area (LPWA)

Low power wide area (LPWA), in licensed spectrum, is an optimal solution for IoT applications that require low-power, low-cost, and low-bandwidth communication. In a smart city, LPWA networks are used in applications such as building automation and industrial automation.

1.4 Wireless Telecommunications

3G, 4G and 5G are cellular wireless telecommunications standards that have been primarily used by mobile phones and data terminals[3]. Importantly, 5G offers lower latency, increased speed, higher density (number of connected devices), added capacity (network throughput), and energy efficiency, as compared to prior wireless generations.

The higher density offered by 5G means that it has the ability to support 10x more connected devices per square kilometer of network, as compared to 4G. This feature is particularly relevant to the proliferation[4] of IoT because a 5G network enables the simultaneous operation of 1 million connected devices in one square kilometer. In a smart city, 5G and IoT become essential for applications such as traffic management, emergency response, and self-driving cars.

1.5 Wireless Sensor Network (WSN)

A wireless sensor network (WSN) is a network that is untethered to any wires and comprises a large number of small, low-power devices called sensor nodes. Specifically, these nodes are equipped with sensors, microcontrollers, and wireless communication capabilities, and are deployed in a variety of environments to monitor and collect data.

In a smart city, a WSN can be used to monitor temperature, humidity, and air quality in a building.

1.6 DASH7

DASH7 is a long-range, low-power wireless communications standard, ideally suited for

---

1 bidirectional short-distance communication：双向短距离通信
2 contactless payment：非接触支付
3 terminal ['tɜːmɪnl] n. 终端
4 proliferation [prəˌlɪfəˈreɪʃn] n. 增殖

sensor networks, active RFID tags, and other IoT devices. This standard is typically used in applications requiring modest bandwidth like text messages, sensor readings, asset tracking, or location-based[1] advertising coordinates. More specifically, these use cases are applied to material monitoring, warehouse optimization, and smart meter developments.

1.7 Addressing

Addressing refers to the process of uniquely identifying and addressing IoT devices and other network entities. This is necessary to ensure that data and commands can be properly routed and delivered to the correct devices and systems in a smart city.

2. IoT Applications for Smart Cities

2.1 Smart Urban[2] Mobility

Traffic congestion is one of the key challenges of every city administration. IoT is playing a key role in alleviating traffic congestion[3] by making various types of real-time data available on vehicular movement.

2.1.1 Traffic Monitoring

Smart traffic management solutions are being used to monitor and analyze traffic flows. These systems optimize traffic lights and help prevent roadways from becoming too congested based on time of day or "rush hour[4]" schedules.

2.1.2 Smart Parking

Smart parking applications use cameras and other sensors to help drivers find available parking spaces without continuously circling around crowded city blocks or parking lots. Sensors placed on parking spots transmit data to a server, which delivers information to drivers via mobile phone applications or display boards.

2.1.3 Connected Vehicles

Connected cars and transport services are growing in adoption, with their ability to provide real-time traffic data and faster routes to drivers.

2.2 Urban Sustainability[5]

Our world is transitioning to more environmentally-aware smart cities and IoT technologies are the catalyst[6] for this shift. Beyond existing initiatives—such as switching to energy-efficient LED[7] lighting or creating low-emission zones—smart lighting, smart meters, and smart waste management are examples of urban sustainability.

---

1 location-based [ləʊˈkeɪʃnˈbeɪst] *adj.* 基于位置的
2 urban [ˈɜːbən] *adj.* 都市的
3 congestion [kənˈdʒestʃən] *n.* 拥挤，堵车
4 rush hour：（上、下班的）高峰期
5 sustainability [səˌsteɪnəˈbɪlɪti] *n.* 持续性
6 catalyst [ˈkætəlɪst] *n.* 催化剂
7 LED (Light Emitting Diode)：发光二极管

2.2.1 Smart Lighting

Smart lighting changes the intensity[1] of street lights based on movement of vehicles and pedestrians. This results in notable energy savings and reduction of light pollution. Also, installing sensors to detect malfunctioning[2] public lights reduces maintenance costs.

2.2.2 Smart Meters

Smart meters are IoT devices that are attached to buildings and connected to a smart energy grid, allowing utility companies to manage energy flow more effectively. Also, smart meters enable users to track their power consumption, leading to more energy usage awareness and potential savings.

2.2.3 Smart Waste Management

Smart waste management can improve efficiency and reduce costs by using capacity sensors to track the level of waste held in garbage cans and recycling containers, determining the most efficient pick-up routes for waste management companies or public services.

2.3 Smart Buildings and Environment

Smart city and IoT technology are also working in tandem to solve problems in air quality, building automation, and noise.

2.3.1 Air Quality Monitoring

Air quality data is being used in cities around the world to support urban planning decisions, such as where to locate new buildings and roads, and to develop and enforce air pollution regulations. Specifically, air quality monitoring has been made possible with optical, electrochemical, and beta attenuation[3] sensors placed around a smart city.

2.3.2 Building Automation

IoT technologies are helping to improve the efficiency, safety, and comfort of public buildings such as schools, libraries, government facilities, and community centers through automation. The goal with building automation is to enhance the end user's experience and reduce operating costs while providing a more sustainable environment.

2.3.3 Noise Monitoring

Different types of sensors and devices can be used for noise monitoring, such as microphones. These IoT devices are placed strategically around a smart city to capture data on noise levels, which is then transmitted to a central monitoring system.

Data from IoT devices allow smart city officials to identify hotspots of noise pollution and bring interventions such as sound barriers, green spaces and noise-reduction building materials.

1　intensity [ɪnˈtensəti] *n.* 亮度
2　malfunction [ˌmælˈfʌŋkʃn] *n.* 故障，失灵
3　attenuation [əˌtenjuˈeɪʃn] *n.* 衰减

参考译文

Text A 无线传感器网络

无线传感器网络由不同位置的自治传感器组成,这些传感器监控物理或环境状况(如温度、声音及压力等)并把其数据通过网络传给主位置。更现代的网络是双向的,它们也能控制传感器的行为。无线传感器网络最早用于像战场监控这样的军事活动;当今这类网络也在许多行业得以应用,如工业过程监控、机器健康监控等。

无线传感器网络由"节点"构成——从数个到成百上千的节点,每个节点连接一个(有时几个)传感器。这种传感器网络节点通常有几部分:带有内置天线或连接了外置天线的无线收发器、一个微型处理器、与传感器连接的电路及能源。能源常常是电池或内嵌的能量收集器。传感器节点也有多种大小,从鞋盒到粉尘不等,尽管必须制造出"尘埃"大小的功能部件。传感器节点的成本也有多种,从几美元到几百美元,这取决于每个节点的复杂性。传感器节点的尺寸和成本限制了资源,如能量、内存、计算速度和通信带宽。无线传感器网络的拓扑结构也不尽相同,从简单的星状网络到高级的多跳无线网状网络。网络之间切换的传输技术可以使用路由法或者泛洪法。

在计算机科学和通信学中,无线传感器网络是活跃的研究领域,每年都有许多研讨会。

1. 特点

无线传感器网络的主要特点包括:

- 能源消耗约束了使用电池或能源收集器的节点;
- 能够应付节点故障;
- 节点具有可移动性;
- 通信故障;
- 异种节点;
- 可扩展到大规模部署;
- 能够抵御严酷的环境条件;
- 易用;
- 能源消耗。

可以把传感器节点想象为小计算机,具有非常简单的接口和部件。通常它由以下部分组成:计算能力有限的处理器、有限的内存、传感器和 MEMS(包括特制电路)、通信设备(通常是无线或光学收发器)及电池形式的电源。它也可能包括能源收集模块、辅助 ASIC 和可能的辅助通信设备(如 RS-232 或 USB)。

基站是一个或多个 WSN 部件,带有更多的计算、能源和通信资源。它们作为传感器节点和终端用户之间的通道,通常把数据从 WSN 转发给服务器。另外的特定部件是基于网络路径的路由器,供运算、计算和分布路由表使用。

2. 平台

2.1 标准和规范

WSN 当前的几个标准都是由包括 WAVE2M 这样的组织发布和开发的。在 WSN 领域有许多标准组织。IEEE 着重于物理和 MAC 层；因特网工程工作小组注重第三层及以上层。除此之外，还有像国际自动化学会这样的团体提供了垂直解决方案，覆盖了各层协议。最终，还有一些非标准的私有机制和规范。

可以用于 WSN 的标准远远少于其他计算系统，这使得大多数系统无法在不同系统之间直接通信。WSN 通信中使用的主要标准包括：
- WirelessHART；
- IEEE 1451；
- ZigBee / 802.15.4；
- ZigBee IP；
- 6LoWPAN。

2.2 硬件

WSN 的一个主要挑战是生产低成本和微小传感器节点。生产 WSN 硬件的小公司日益增多，这与 20 世纪 70 年代家庭计算的商业境况颇为相似。许多节点仍然处于研发期，尤其是软件。传感器网络固有地使用极低功率的方式收集数据。

2.3 软件

能量是 WSN 节点最稀缺的资源，这也决定了 WSN 的寿命。要打算在各种环境（包括远程和野外）中部署 WSN，通信是关键。因此，算法和协议必须解决以下问题：寿命最长、稳健性、容错和自配置。

因为有限的能源决定了寿命，所以敏感元件的能源消耗必须最小并且传感器更有能效。在节点不工作时必须关闭无线电源以便节能。

在 WSN 软件研究中一些重要的话题是操作系统、安全性和移动性。

用于无线传感器网络节点的操作系统通常比一般的操作系统简单。有两个理由使其更像嵌入式系统。首先，无线传感器网络通常为特定应用而部署，而不是在普通平台上；其次，低成本和低功率需求使得大多数无线传感器节点都有低功率微控制器，这就使这些装置（如虚拟内存）要么不必要安装，要么安装成本太贵。

因此，传感器网络可能使用像 eCos 或 μC/OS 这样的嵌入式操作系统。但是，这样的操作系统通常具有实时操作的特点。

TinyOS 或许是第一个专门为无线传感器网络设计的操作系统。TinyOS 基于事件驱动编程模型而不是多线程模型。TinyOS 程序由事件处理器和执行到底的任务组成。当一个外部事件发生时，如进来一个数据包或传感器读入了数据，TinyOS 就发送信号给适当的事件处理器来处理该事件。随后，事件处理器可以由 TinyOS 内核预先安排任务。

LiteOS 是一个新开发的无线传感器网络操作系统，它提供类似 UNIX 的概念并支持 C 语言编程。

Contiki 是使用 C 语言编程风格比较简单的操作系统，它同时也提供像 6LoWPAN 和 Protothreads 这样的先进技术。

3. 其他概念

3.1 分布式传感器网络

如果在传感器网络中使用集中结构并且中心节点出了故障，整个网络就会崩溃，但是可以使用分布式控制结构来增加这种传感器网络的可靠性。

在 WSN 中使用分布式控制的理由如下：
- 传感器节点容易出故障；
- 更好地收集数据；
- 在中心节点出故障时提供备份；
- 没有核心的实体来分配资源并且它们必须是自组织的。

3.2 数据集成和传感器网络

从无线传感器网络收集的数据通常以数字格式存储在中心基站。另外，OGC 指定了专门用于互用接口和元数据编码的标准，使用这些标准能够实时地把异种传感器网络整合到因特网中，这样任何人都可以通过网络浏览器监控无线传感器网络。

3.3 网内进程

为减少通信成本，有些算法删除或减少节点冗余传感器信息并避免转发无用信息。因为节点可以检查它们转发的信息，所以它们可以测量来自其他节点的读数的平均值或方向。例如，在感知和监控应用中，监控环境的相邻的传感器节点通常注册类似的数值。这种数据冗余来自传感器的空间关系，它激发了网内数据收集和挖掘技术。

Unit 7

Text A

How Wireless Networks Work

A wireless network or wireless local area network (WLAN) serves the same purpose as a wired one — to link a group of computers. Because "wireless" doesn't require costly wiring, the main benefit is that it's generally easier, faster and cheaper to set up.

By comparison, creating a network by pulling wires throughout the walls and ceilings of an office can be labor-intensive and thus expensive. But even when you have a wired network already in place, a wireless network can be a cost-effective way to expand or augment it. In fact, there's really no such thing as a purely wireless network, because most of the wireless networks link back to a wired network at some point.

1. The Basics

Wireless networks operate using radio frequency (RF) technology, a frequency within the electromagnetic spectrum associated with radio wave propagation. When an RF current is supplied to an antenna, an electromagnetic field is created that then is able to propagate through space.

The cornerstone of a wireless network is a device known as an access point (AP[1]). The primary job of an access point is to broadcast a wireless signal that computers can detect and "tune" into. Since wireless networks are usually connected to wired ones, an access point also often serves as a link to the resources available on the a wired network, such as an Internet connection.

---

1 Short for access point, a hardware device or a computer's software that acts as a communication hub for users of a wireless device to connect to a wired LAN. APs are important for providing heightened wireless security and for extending the physical range of service a wireless user has access to.

In order to connect to an access point and join a wireless network, computers must be equipped with wireless network adapters[1]. These are often built right into the computer, but if not, just about any computer or notebook can be made wireless-capable through the use of an add-on adapter[2] plugged into an empty expansion slot, USB port, or in the case of notebooks, a PC Card slot.

2. Wireless Technology Standards

Because there are multiple technology standards for wireless networking, it pays to do your homework before buying any equipment. The most common wireless technology standards include the following:

- 802.11b: the first widely used wireless networking technology, known as 802.11b (more commonly called WiFi), first debuted almost a decade ago, but is still in use.
- 802.11g: in 2003, a follow-on version called 802.11g appeared offering greater performance (that is, speed and range) and remains today's most common wireless networking technology.
- 802.11n: released in 2009, it provides faster wireless transmission speeds and better coverage range. Currently, the 802.11n standard has been replaced by more advanced wireless standards such as 802.11ac (WiFi 5) and 802.11ax (WiFi 6). These new standards offer higher speeds, lower latency, and better performance. However, due to the lower cost of the 802.11n standard and the fact that many devices still support it, there are still many networks that use 802.11n for communication. For some low-demand applications, such as home networks or small enterprise networks, 802.11n remains a viable choice.

All the WiFi variants (802.11b, g and n products) use the same 2.4 GHz radio frequency, and as a result are designed to be compatible with each other, so you can usually use devices based on the different standards within the same wireless network. The catch is that doing so often requires special configuration to accommodate the earlier devices, which in turn can reduce the overall performance of the network. In an ideal scenario you'll want all your wireless devices, the access point and all wireless-capable computers to be using the same technology standard and to be from the same vendor whenever possible.

3. Wireless Speed & Range

When you buy a piece of wireless network hardware, it will often quote performance figures (i.e., how fast it can transmit data) based on the type of wireless networking standard it uses, plus any added technological enhancements. In truth, these performance figures are almost

---

1 Often abbreviated as NIC, an expansion board you insert into a computer so the computer can be connected to a network. Most NICs are designed for a particular type of network, protocol, and media, although some can serve multiple networks.

2 The circuitry required to support a particular device. For example, video adapters enable the computer to support graphics monitors, and network adapters enable a computer to attach to a network. Adapters can be built into the main circuitry of a computer or they can be separate add-ons that come in the form of expansion boards.

always wildly optimistic.

While the official speeds of 802.11b, 802.11g, and 802.11n networks are 11, 54, and 270 megabits per second (Mbps) respectively, these figures represent a scenario that is simply not attainable in the real world. As a general rule, you should assume that in a best-case scenario you'll get roughly one-third of the advertised performance.

It's also worth noting that a wireless network is by definition a shared network, so the more computers you have connected to a wireless access point, the less data each will be able to send and receive. Just as a wireless network's speed can vary greatly, the range can change too. For example, 802.11b and g officially work over a distance of up to 100 meters indoors or 300 meters outdoors, but the key term there is "up to". Chances are you won't see anywhere close to those numbers.

As you might expect, the closer you are to an access point, the stronger the signal and the faster the connection speed. The range and speed you get out of wireless network will also depend on the kind of environment in which it operates. And that brings us to the subject of interference.

4. Wireless Interference

Interference is an issue with any form of radio communication, and a wireless network is no exception. The potential for interference is especially great indoors, where different types of building materials (concrete, wood, drywall, metal, glass and so on) can absorb or reflect radio waves, affecting the strength and consistency of a wireless network's signal. Similarly, devices like microwave ovens and some cordless phones can cause interference because they operate in the same 2.4 GHz frequency range as 802.11b/g/n networks. You can't avoid interference entirely, but in most cases it's not significant enough to affect the usability of the network. When it does, you can usually minimize the interference by relocating wireless networking hardware or using specialized antennas[1].

5. Data Security on Wireless Networks

In the same way that all you need to pick up a local radio station is a radio, all anyone needs to detect a wireless network within nearby range is a wireless-equipped computer. There's no way to selectively hide the presence of your network from strangers, but you can prevent unauthorized people from connecting to it, and you can protect the data traveling across the network from prying eyes. By turning on a wireless network's encryption feature, you can scramble the data and control access to the network.

Wireless network hardware supports several standard encryption schemes, but the most

---

1 Also called an aerial (['eərɪəl] *n.* 天线), an antenna is a conductor that can transmit, send and receive signals such as microwave, radio or satellite signals. A high-gain antenna increases signal strength, where a low-gain antenna receives or transmits over a wide angle.

common are wired equivalent privacy (WEP[1]), WiFi protected access (WPA[2]), and WiFi protected access 2 (WPA2[3]). WEP is the oldest and least secure method and should be avoided.

WPA is a relatively more secure choice, and WPA2 is widely regarded as the most secure protocol. For enhancing the security of wireless networks, using WPA2 is the best choice. Unless you intend to provide public access to your wireless network — and put your business data or your own personal data at risk — you should consider encryption mandatory.

New Words

ceiling	['siːlɪŋ]	n. 天花板，最高限度
labor-intensive	['leɪbən'tensɪv]	adj. 劳动密集型的
cost-effective	[ˌkɒstɪ'fektɪv]	adj. 有成本效益的，划算的
augment	[ɔːg'ment]	v. 增加，增大
		n. 增加
electromagnetic	[ɪˌlektrəʊmæg'netɪk]	adj. 电磁的
spectrum	['spektrəm]	n. 光谱，频谱
cornerstone	['kɔːnəstəʊn]	n. 墙角石，基础
tune	[tjuːn]	vt. 收听
adapter	[ə'dæptə]	n. 适配器
debut	['deɪbjuː]	v. 出现，亮相
finalize	['faɪnəlaɪz]	v. 把（计划、稿件等）最后定下来，定案
draft	[drɑːft]	n. 草稿，草案
upgrade	['ʌpgreɪd]	n. 升级
	[ˌʌp'greɪd]	vt. 使升级
catch	[kætʃ]	n. 捕捉
		v. 捕获
		vi. 抓住
accommodate	[ə'kɒmədeɪt]	vt. 供给，使适应，调节，调和
		vi. 适应
reduce	[rɪ'djuːs]	vt. 减少，缩小，简化，还原

1　A security protocol for wireless local area networks (WLANs) defined in the 802.11b standard. WEP is designed to provide the same level of security as that of a wired LAN. LANs are inherently more secure than WLANs because LANs are somewhat protected, having some or all part of the network inside a building that can be protected from <u>unauthorized access</u>（未授权的访问）.

2　A WiFi standard that was designed to improve upon the security features of WEP. The technology is designed to work with existing WiFi products that have been enabled with WEP (i.e., as a software upgrade to existing hardware).

3　Short for WiFi protected access 2, the follow on security method to WPA for wireless networks that provides stronger data protection and network access control. It provides enterprise and consumer WiFi users with a high level of assurance that only authorized users can access their wireless networks. Based on the IEEE 802.11i standard, WPA2 provides government grade security by implementing the <u>National Institute of Standards and Technology</u> (NIST，美国国家标准及技术协会) <u>FIPS</u> (Federal Information Processing Standards，美国联邦信息处理标准) 140-2 compliant <u>AES</u> (Advanced Encryption Standard，高级加密标准) encryption algorithm and 802.1x-based authentication.

scenario	[sə'nɑːrɪəʊ]	n. 情景
quote	[kwəʊt]	vt. 提供，提出，报（价）
figure	['fɪgə]	n. 外形，轮廓，图形，画像，数字，形状
		vt. 描绘，表示，象征
optimistic	[ˌɒptɪ'mɪstɪk]	adj. 乐观的
respectively	[rɪ'spektɪvli]	adv. 分别地，各个地
attainable	[ə'teɪnəbl]	adj. 可到达的，可得到的
assume	[ə'sjuːm]	vt. 假定，设想
roughly	['rʌflɪ]	adv. 概略地，粗糙地
potential	[pə'tenʃl]	adj. 潜在的，可能的
concrete	['kɒŋkriːt]	n. 混凝土
drywall	['draɪwɔːl]	n.（不抹灰的）板墙，干墙，石膏板预制件
reflect	[rɪ'flekt]	v. 反射
consistency	[kən'sɪstənsi]	n. 密度，一致性，连贯性
cordless	['kɔːdləs]	n. 不用电线的
entirely	[ɪn'taɪəli]	adv. 完全地，全然地，一概地
usability	[ˌjuːzə'bɪlɪti]	n. 可用性
relocate	[ˌriːləʊ'keɪt]	v. 重新部署
prying	['praɪɪŋ]	adj. 爱打听的
		v. 打听，刺探（他人的私事）
encryption	[ɪn'krɪpʃn]	n. 编密码，加密
mandatory	['mændətəri]	adj. 命令的，强制的，托管的

Phrases

by comparison	比较起来
pull wire	拉线
wired network	有线网络
be associated with	与……有联系，与……有关
electromagnetic field	电磁场
electromagnetic spectrum	电磁波频谱
be equipped with	装备
network adapter	网络适配器，网卡
just about	几乎
plug into	把（电器）插头插入，接通
expansion slot	扩充插槽
in the case of	在……的情况
be compatible with	适合，一致
overall performance	总性能，全部工作特性

a piece of	一套，一件
in truth	实际上
performance figure	性能指标
cordless phone	无绳电话
prevent sb. from doing sth.	阻止某人做某事

Abbreviations

WLAN (Wireless Local Area Network)	无线局域网
RF (Radio Frequency)	无线电频率
AP (Access Point)	访问接入点
Mbps (Megabits per second)	兆位每秒
WEP (Wired Equivalent Privacy)	有线等效加密
WPA (WiFi Protected Access)	WiFi 保护接入

Analysis of Difficult Sentences

[1] The primary job of an access point is to broadcast a wireless signal that computers can detect and "tune" into.

本句中，to broadcast a wireless signal that computers can detect and "tune" into 是一个动词不定式短语，作表语。在该短语中，that computers can detect and "tune" into 是一个定语从句，修饰和限定 a wireless signal。

[2] Since wireless networks are usually connected to wired ones, an access point also often serves as a link to the resources available on the a wired network, such as an Internet connection.

本句中，Since wireless networks are usually connected to wired ones 是一个原因状语从句，修饰谓语 serves as。such as an Internet connection 是对 a link to the resources available on the a wired network 的举例说明。

[3] Because there are multiple technology standards for wireless networking, it pays to do your homework before buying any equipment.

本句中，Because there are multiple technology standards for wireless networking 是一个原因状语从句，修饰谓语 pays。it 是形式主语，真正的主语是动词不定式短语 to do your homework before buying any equipment。

[4] While the official speeds of 802.11b, 802.11g, and 802.11n networks are 11, 54, and 270 megabits per second (Mbps) respectively, these figures represent a scenario that is simply not attainable in the real world.

本句中，While the official speeds of 802.11b, 802.11g, and 802.11n networks are 11, 54, and 270 megabits per second (Mbps) respectively 是一个让步状语从句，修饰谓语 represent。that is simply not attainable in the real world 是一个定语从句，修饰和限定宾语 a scenario。

[5] Unless you intend to provide public access to your wireless network — and put your business data or your own personal data at risk — you should consider encryption mandatory.

本句中，Unless you intend to provide public access to your wireless network — and put your business data or your own personal data at risk 是一个条件状语从句，修饰谓语 should consider。在该从句中，and put your business data or your own personal data at risk 是对 provide public access to your wireless network 的补充说明，说明这样做的结果。Unless 的意思是"除非，如果不"。

Exercises

【EX.1】Answer the following questions according to the text.

1. What is the purpose of a wireless network?
2. What is the cornerstone of a wireless network? What is its primary job?
3. What must computers be equipped with in order to connect to an access point and join a wireless network?
4. What do the most common wireless technology standards include?
5. What radio frequency do all the WiFi variants (802.11b, g and n products) use?
6. What are the official speeds of 802.11b, 802.11g, and 802.11n networks?
7. What do the range and speed you get out of wireless network depend on?
8. Why can devices like microwave ovens and some cordless phones cause interference?
9. What are the most common standard encryption schemes wireless network hardware supports?
10. Is there really such thing as a purely wireless network in fact? Why?

【EX.2】Translate the following terms or phrases from English into Chinese and vice versa.

1.	wired network	1.	
2.	electromagnetic field	2.	
3.	electromagnetic spectrum	3.	
4.	network adapter	4.	
5.	be equipped with	5.	
6.	overall performance	6.	
7.	performance figure	7.	
8.	expansion slot	8.	
9.	spectrum	9.	
10.	adapter	10.	
11.	n. 升级　vt. 使升级	11.	
12.	v. 反射	12.	
13.	v. 重新部署	13.	
14.	n. 可用性	14.	
15.	n. 编密码，加密	15.	

【EX.3】 **Translate the following sentences into Chinese.**
1. Electromagnetic waves travel at the same speed as light.
2. An example of a physical interface is a network adapter.
3. Engineers will test the performance of the computer.
4. The Internet is sometimes described as a cloud—a big cordless (borderless) area of computing power.
5. The database security involves data secret, integrity and usability.
6. Key management is a critical technic of realizing database encryption.
7. It mainly includes symmetric encryption algorithms and asymmetric cryptographic algorithms and protocols.
8. The authorities are considering implementing a mandatory electronic recycling program.
9. You may not connect your laptop to the wired network in the computer labs.
10. Select this option to provide authenticated network access for wired and wireless Ethernet networks.

【EX.4】 **Complete the following passage with appropriate words in the box.**

close	reader	simultaneously	indoor	operate
require	single	positioned	attached	needed

RFID chips are quite similar to bar code labels in that they typically work with a corresponding scanner or reader. However, RFID chips have significant advantages. Because an RFID chip communicates with a ___1___ through radio waves (not infrared, which is being used by bar code technology), the chip doesn't have to be ___2___ right in front of the reader. That is, line-of-sight is not ___3___.

Also, unlike a bar code reader/label pair, which have to be really ___4___ (about a few centimeters), some RFID reader/chip pairs can function even if they are a few meters apart. Furthermore, while a bar code label can only be read by a ___5___ reader at a time, an RFID chip can transmit data to multiple readers ___6___.

There are different kinds of RFID chips. Some ___7___ batteries, known as "active" chips, while others don't, known as "passive". Others are designed for ___8___ use, while others are built for rugged, outdoor applications. The most common applications include object tracking and identification.

Chips can also differ in the kind of radio frequencies they ___9___ on. Some communicate via UHF (Ultra High Frequency), others HF (High Frequency), and still others LF (Low Frequency).

RFID chips can be ___10___ just about anywhere: clothes, shoes, vehicles, containers, and even plants, animals, and human beings (as implants). Miniaturized chips have even been attached to insects.

【EX.5】Translate the following passage into Chinese.

Radio Frequency Identification Tag (RFID Tag)

Although RFID tags have similar applications to barcodes, they are far more advanced. For instance, reading information from an RFID tag does not require line-of-sight and can be performed over a distance of a few meters. This also means that a single tag can serve multiple readers at a time, compared to only one for a bar code tag.

In the context of RFID technology, the term "tag" is also meant to include labels and cards. The kind of tag depends on the body or object on which the tag will be attached to. RFID systems can operate either in UHF (Ultra High Frequency), HF (High Frequency), or LF (Low Frequency). Thus, tags can also vary in terms of the frequencies on which they will operate.

These tags can be attached to almost any object. Although the usual target objects are apparel, baggage, containers, construction materials, laundry, and bottles, they have also been attached to animals, humans, and vehicles.

Some RFID tags are designed for rugged, outdoor-based applications. These are built to endure natural and incandescent light, vibration, shock, rain, dust, oil, and other harsh conditions. They are normally passive, i.e., they don't require batteries to function. Thus, they can operate 24/7 without risk of losing power. Such heavy-duty tags are usually attached to trucks, cargo containers, and light rail cars for cargo tracking, fleet management, vehicle tracking, and vehicle identification, among others.

Text B

RFID

Radio frequency identification (RFID) is the use of a wireless noncontact system that uses radio-frequency electromagnetic fields to transfer data from a tag attached to an object for the purposes of automatic identification and tracking. Some tags require no battery and are powered by the electromagnetic fields used to read them. Others use a local power source and emit radio waves (electromagnetic radiation[1] at radio frequencies). The tag contains electronically stored information which can be read from up to several meters (yards) away. Unlike a bar code, the tag does not need to be within line of sight of the reader and may be embedded in the tracked object.

RFID tags are used in many industries. An RFID tag attached to an automobile during production can be used to track its progress through the assembly line. Pharmaceuticals can be tracked through warehouses. Livestock and pets may have tags injected[2], allowing positive identification of the animal.

1　Electromagnetic radiation (EM radiation or EMR) is a form of energy emitted and absorbed by charged particles, which exhibits wave-like behavior as it travels through space.

2　A microchip implant is an identifying integrated circuit placed under the skin of a dog, cat, horse, parrot or other animal. The chip, about the size of a large grain of rice, uses passive RFID (Radio Frequency Identification) technology.

Since RFID tags can be attached to clothing, possessions, or even implanted within people[1], the possibility of reading personally-linked information without consent has raised privacy concerns.

1. Design of RFID

A radio frequency identification system uses tags, or labels attached to the objects to be identified. Two-way radio transmitter-receivers called interrogators or readers send a signal to the tag and read its response. The readers generally transmit their observations to a computer system running RFID software or RFID middleware.

The tag's information is stored electronically in a nonvolatile memory. The RFID tag includes a small RF transmitter and receiver. An RFID reader transmits an encoded radio signal to interrogate the tag. The tag receives the message and responds with its identification information. This may be only a unique tag serial number, or may be product-related information such as a stock number, lot or batch number, production date, or other specific information.

RFID tags can be either passive, active or battery assisted passive. An active tag has an on-board battery and periodically transmits its ID signal. A battery assisted passive (BAP) has a small battery on board and is activated when in the presence of a RFID reader. A passive tag is cheaper and smaller because it has no battery. Instead, the tag uses the radio energy transmitted by the reader as its energy source. The interrogator must be close for RF field to be strong enough to transfer sufficient power to the tag. Since tags have individual serial numbers, the RFID system design can discriminate several tags that might be within the range of the RFID reader and read them simultaneously.

Tags may either be readonly, having a factory-assigned serial number that is used as a key into a database, or may be read/write, where object-specific data can be written into the tag by the system user. Field programmable tags may be write-once, read-multiple; "blank" tags may be written with an electronic product code by the user.

RFID tags contain at least two parts: an integrated circuit for storing and processing information, modulating[2] and demodulating[3] a radio frequency (RF) signal, collecting DC power from the incident reader signal, and other specialized functions; and an antenna for receiving and transmitting the signal.

Fixed readers are set up to create a specific interrogation zone which can be tightly controlled. This allows a highly defined reading area for when tags go in and out of the

---

1 A human microchip implant is an integrated circuit device or RFID transponder encased in silicate glass（硅酸盐玻璃）and implanted in the body of a human being. A subdermal implant typically contains a unique ID number that can be linked to information contained in an external database, such as personal identification, medical history, medications, allergies（['ælədʒi] n. 敏感症）, and contact information.

2 In electronics and telecommunications, modulation is the process of varying one or more properties of a high-frequency periodic waveform（['weɪvfɔːm] n. 波形）, called the carrier signal（载波信号）, with a modulating signal which typically contains information to be transmitted.

3 Demodulation is the act of extracting the original information-bearing signal from a modulated carrier wave.

interrogation zone. Mobile readers may be hand-held or mounted on carts or vehicles.

Signaling between the reader and the tag is done in several different incompatible ways, depending on the frequency band used by the tag. Tags operating on LF and HF frequencies are, in terms of radio wavelength, very close to the reader antenna, only a small percentage of a wavelength away. In this near field[1] region, the tag is closely coupled electrically with the transmitter in the reader. The tag can modulate the field produced by the reader by changing the electrical loading the tag represents. By switching between lower and higher relative loads, the tag produces a change that the reader can detect. At UHF and higher frequencies, the tag is more than one radio wavelength away from the reader, requiring a different approach. The tag can backscatter[2] a signal. Active tags may contain functionally separated transmitters and receivers, and the tag need not respond on a frequency related to the reader's interrogation signal.

An electronic product code (EPC) is one common type of data stored in a tag. When written into the tag by an RFID printer, the tag contains a 96-bit string of data. The first eight bits are a header which identifies the version of the protocol. The next 28 bits identify the organization that manages the data for this tag; the organization number is assigned by the EPCGlobal consortium. The next 24 bits are an object class, identifying the kind of product; the last 36 bits are a unique serial number for a particular tag. These last two fields are set by the organization that issued the tag. Rather like a URL, the total electronic product code number can be used as a key into a global database to uniquely identify a particular product.

Often more than one tag will respond to a tag reader, for example, many individual products with tags may be shipped in a common box or on a common pallet. Collision detection is important to allow reading of data. Two different types of protocols are used to "singulate" a particular tag, allowing its data to be read in the midst of many similar tags. In a slotted Aloha system, the reader broadcasts an initialization command and a parameter that the tags individually use to pseudorandomly delay their responses. When using an "adaptive binary tree" protocol, the reader sends an initialization symbol and then transmits one bit of ID data at a time; only tags with matching bits respond, and eventually only one tag matches the complete ID string.

Both methods have drawbacks when used with many tags or with multiple overlapping readers.

2. Miniaturization

RFIDs are easy to conceal or incorporate in other items. For example, in 2009 researchers at Bristol University successfully glued RFID micro-transponders to live ants in order to study their behavior. This trend towards increasingly miniaturized RFIDs is likely to continue as technology

---

1 The near field (or near-field) and far field (or far-field) and the transition zone are regions of time varying electromagnetic field around any object that serves as a source for the field.

2 In physics, backscatter (or backscattering) is the reflection of waves, <u>particles</u>（['pɑːtɪkl] *n.* 粒子）, or signals back to the direction from which they came. It is a diffuse reflection due to scattering, as opposed to <u>specular reflection</u>（镜面反射）like a mirror.

advances.

Hitachi holds the record for the smallest RFID chip, at 0.05mm × 0.05mm. This is 1/64th the size of the previous record holder, the mu-chip. Manufacture is enabled by using the silicon on insulator[1] (SOI) process. These dust-sized chips can store 38-digit numbers using 128-bit read only memory (ROM[2]). A major challenge is the attachment of the antennas, thus limiting read range to only millimeters.

New Words

noncontact	[ˈnɒnˈkɒntækt]	*adj.* 没有接触的，非接触的
tag	[tæg]	*n.* 标签
		vt. 加标签于
track	[træk]	*n.* 轨迹，跟踪
		vt. 追踪
emit	[ɪˈmɪt]	*vt.* 发出，放射
yard	[jɑːd]	*n.* 码（长度单位）
reader	[ˈriːdə]	*n.* 阅读器，读卡机
livestock	[ˈlaɪvstɒk]	*n.* 家畜，牲畜
inject	[ɪnˈdʒekt]	*vt.* 引入，插入，注入
implant	[ɪmˈplɑːnt]	*v.* 植入，插入
label	[ˈleɪbl]	*n.* 标签，签条，标志
		vt. 贴标签于
identify	[aɪˈdentɪfaɪ]	*vt.* 识别，鉴别
interrogator	[ɪnˈterəʊgeɪtə]	*n.* 询问器，询问机，侦测器
middleware	[ˈmɪdlweə]	*n.* 中间设备，中间件
nonvolatile	[nɒnˈvɒlətaɪl]	*adj.* 非易失性的
memory	[ˈmeməri]	*n.* 存储器，内存
encode	[ɪnˈkəʊd]	*vt.* 编码，把（电文、情报等）译成电码（或密码）
interrogate	[ɪnˈterəgeɪt]	*vt.* 询问
active	[ˈæktɪv]	*adj.* 有源的
periodically	[ˌpɪərɪˈɒdɪkəli]	*adv.* 周期性地，定时性地
discriminate	[dɪˈskrɪmɪneɪt]	*v.* 区别，区别待遇
programmable	[ˈprəʊgræməbl]	*adj.* 可设计的，可编程的
modulate	[ˈmɒdjuleɪt]	*vt.* （信号）调制

1　Silicon on insulator (SOI) technology refers to the use of a layered silicon-insulator-silicon <u>substrate</u>（[ˈsʌbstreɪt] *n.* 底层,下层）in place of conventional silicon substrates in semiconductor manufacturing, especially <u>microelectronics</u>（[ˌmaɪkrəʊɪˌlekˈtrɒnɪks] *n.* 微电子学）, to reduce <u>parasitic</u>（[ˌpærəˈsɪtɪk] *adj.* 寄生的）device capacitance, thereby improving performance.

2　Read only memory (ROM) is a class of storage medium used in computers and other electronic devices. Data stored in ROM cannot be modified, or can be modified only slowly or with difficulty, so it is mainly used to distribute firmware (software that is very closely tied to specific hardware, and unlikely to need frequent updates).

demodulate	[diːˈmɒdjʊleɪt]	vt.	解调
incident	[ˈɪnsɪdənt]	n.	入射
tightly	[ˈtaɪtli]	adv.	紧紧地，坚固地
hand-held	[hændheld]	adj.	手持的
incompatible	[ˌɪnkəmˈpætəbl]	adj.	不兼容的
band	[bænd]	n.	波段
wavelength	[ˈweɪvleŋθ]	n.	波长
approach	[əˈprəʊtʃ]	n.	方法，步骤，途径
backscatter	[ˈbækˌskætə]	n.	反向散射，背反射
pallet	[ˈpælət]	n.	货盘
singulate	[ˈsɪŋgjuleɪt]	vt.	挑出
pseudorandom	[psjuːdəʊˈrændəm]	adj.	伪随机的
initialization	[ɪˌnɪʃəlaɪˈzeɪʃn]	n.	设定初值，初始化
miniaturization	[ˌmɪnətʃəraɪˈzeɪʃn]	n.	小型化
transponder	[trænˈspɒndə]	n.	发射机应答器，询问机，转发器
miniaturize	[ˈmɪnətʃəraɪz]	vt.	使小型化

Phrases

local power source	本地电源
electromagnetic radiation	电磁辐射
radio frequency	无线电频率
bar code	条形码
line of sight	视线，瞄准线
assembly line	（工厂产品的）装配线
radio signal	无线电信号
serial number	序号，序列号
stock number	物料编号
batch number	批号，批数
on-board battery	板载电池
in the presence of	在面前
read only	只读
write-once, read-multiple	单次写入多次读取（WORM）
integrated circuit	集成电路
interrogation zone	读取器询问区，侦测区
near field	近场
in the midst of	在……之中，在……的中途
binary tree	二叉树

Abbreviations

ID (Identification, Identity)　　　　身份
BAP (Battery Assisted Passive)　　　电池辅助无源
DC (Direct Current)　　　　　　　　直流电
LF (Low Frequency)　　　　　　　　低频
HF (High Frequency)　　　　　　　　高频
EPC (Electronic Product Code)　　　 电子产品代码
EPCGlobal　　　　　　　　　　　　国际物品编码协会EAN和美国统一代码委员会（UCC）的一个合资公司
SOI (Silicon On Insulator)　　　　　　绝缘体上硅薄膜
ROM (Read Only Memory)　　　　　 只读存储器

Exercises

【EX.6】Fill in the blanks according to the text.

1. Radiofrequency identification (RFID) uses _____ that uses radio-frequency electromagnetic fields to transfer data from a tag attached to an object for the purposes of _____.
2. A radio frequency identification system uses _____ or _____ attached to the objects to be identified.
3. The tag's information is stored electronically in a _____. The RFID tag includes _____.
4. RFID tags can be either _____, _____ or _____. An active tag has _____ and periodically transmits its ID signal.
5. A passive tag is cheaper and smaller because _____. Instead, the tag uses the radio energy transmitted by _____ as its energy source.
6. RFID tags contain at least two parts: _____ for storing and processing information, modulating and demodulating a radio frequency (RF) signal, collecting DC power from the incident reader signal, and other specialized functions; and _____ for receiving and transmitting the signal.
7. An electronic product code (EPC) is _____ stored in a tag. When written into the tag by an RFID printer, the tag contains _____.
8. _____ can be used as a key into a global database to uniquely identify a particular product.
9. In a slotted Aloha system, the reader broadcasts _____ and _____ that the tags individually use to pseudorandomly delay their responses.
10. _____ holds the record for the smallest RFID chip, _____. This is 1/64th the size of the previous record holder, _____.

【EX.7】Translate the following terms or phrases from English into Chinese and vice versa.

1. ____bar code____　　　　　　　　1. _____

2.	radio frequency	2.	
3.	local power source	3.	
4.	integrated circuit	4.	
5.	electromagnetic radiation	5.	
6.	near field	6.	
7.	noncontact	7.	
8.	radio signal	8.	
9.	track	9.	
10.	label	10.	
11.	*vt.* 识别，鉴别	11.	
12.	*n.* 中间设备，中间件	12.	
13.	*adj.* 非易失性的	13.	
14.	*adj.* 无源的	14.	
15.	*adj.* 有源的	15.	

【EX.8】 Translate the following sentences into Chinese.

1. Bar codes have replaced the traditional price tag in big stores.
2. Most often, the IBM disk-track format is used, sometimes with minor variations.
3. They also emit UHF waves of 225~400 MHz.
4. The photo-electric reader is capable of scanning characters at the rate of two thousand a second.
5. The development of data warehouse technology provides advantage for accomplishing this task.
6. Label is a fixed length of 24 characters and contains standard information.
7. The middleware technology is effectively developed for resolving the distributed computing problems.
8. The ROM chip retains instructions in a permanently accessible, nonvolatile form.
9. The programmable controller (PLC) is used at abroad due to its higher reliability and convenient application.
10. This will also change according to the wavelength of the laser light.

Reading Material

Near Field Communication (NFC)

1. What Is Near Field Communication (NFC)?

Near field communication (NFC) is a short-range wireless connectivity technology that uses magnetic field[1] induction to enable communication between devices when they're touched together or brought within a few centimeters of each other. This includes authenticating credit cards, enabling physical access, transferring small files and jumpstarting more capable wireless links. Broadly speaking, it builds on and extends the work of existing ecosystems and standards around radio frequency ID (RFID) tags.

NFC extends RFID and contactless capabilities with more dynamic features enabled by modern smartphones. All modern phones now support NFC chips and applications to take advantage of the billions of RFID tags and terminals already deployed. NFC makes it easier to load multiple cards into a single phone for payments, municipal transit[2], building access, opening car doors and other use cases. NFC supports interactive applications built on basic RFID capabilities such as automatically pairing Bluetooth headphones and WiFi connections. It can also automatically pull up[3] data or an app from a poster or ad[4].

Modern services, such as Google Nearby Share, employ NFC to configure wireless services across faster networks like Bluetooth or WiFi Direct.

NFC is limited to short-range communication, which has important implications for physical access security. A user must be within 3.5 inches (10 cm) of an NFC terminal to process a payment or open a door. Another important aspect is that no power is required for the basic mechanics of listening to and responding to NFC requests. This makes it possible to implement in items that lack a battery, such as credit cards.

NFC also complements wireless technologies such as Bluetooth, Ultrawideband[5] (UWB), WiFi Direct and QR codes[6]. Its most significant advantage is that it is the easiest wireless technology for setting up a connection, which makes it useful for IoT devices. However, it is not as good at maintaining a connection over distances or for long periods.

2. How Does NFC work?

NFC works on top of three crucial innovations in wireless tag readers, cryptographic credit card processing and peer-to-peer (P2P)[7] connectivity to enable various applications.

NFC builds on the work of the RFID set of standards and specifications, such as ISO/IEC 14443 and ISO/IEC 15963. They utilize a wireless communication technology that differs from most radios in terms of physical principles. Whereas most radios transmit data via radio wave propagation, NFC transmits data via magnetic field induction. NFC data transmits data at 13.56 MHz, which corresponds to a wavelength[8] of 22 meters.

---

1 magnetic field：磁场
2 municipal transit：市政交通
3 pull up：提取
4 ad [æd] n. 广告
5 ultrawideband [ˈʌltrəwaɪdbænd] n. 超宽带
6 QR code：二维码
7 peer-to-peer [piətuːpiə] adj. 对等的
8 wavelength [ˈweɪvleŋθ] n. 波长

One critical aspect of transmitting data via induction coupling rather than radio waves is that the field fades out[1] far more quickly than radio waves. This is useful for preventing people from listening in on sensitive conversations about credit card transactions, door access codes or other confidential information.

The second significant NFC innovation involves cryptographic credit card processing used for contactless payments. Public-key cryptography allows the card to generate a new authentication code for each transaction without revealing the raw card details or three-digit code on the back. This ensures that even if someone were to listen in or a hacker queried a card on a busy subway, they would never glean the original card details.

The NFC forum, a nonprofit industry association[2], took these two building blocks and added P2P connectivity on top of the ISO/IEC 18092 standard. Classic RFID and credit card use cases involve an active card reader that queries a passive tag or card, which is a one-way interaction. The NFC forum introduced specifications that allowed more capable devices like smartphones, headphones, routers, home appliances and industrial equipment to initiate or react to NFC queries. This opened a wide range of interaction and connectivity patterns. It also took a lot of work to simplify the exchange of information while minimizing security vulnerabilities.

Smartphone vendors are starting to build some basic application execution capabilities on top of this. In the Google ecosystem, a smart tag might launch a progressive web app running in the browser.

3. Examples of NFC

A few examples of NFC use cases include the following:

- Mobile payments;
- Transit card payments;
- Ticket redemption[3] at a concert or theater;
- Access authentication for doors or offices;
- Unlocking car doors or rental scooters;
- Venue or location check-in to alert friends on social media;
- Device pairing smartphones and headsets by tapping them together;
- Automatic setup for WiFi connectivity by tapping a phone to a router or internet gateway;
- Connection via smartphone to a radiator to configure its temperature settings and schedule; and
- Connection via smartphone or tablet to industrial equipment to access a more complex control panel.

1　fade out：消失，渐弱
2　nonprofit industry association：非营利行业协会
3　ticket redemption：售票

4. Benefits of NFC

NFC has several real-world benefits, including the following:
- Increases operational efficiency for payment processors;
- Ensures more security than traditional credit cards for payments;
- Allows users to choose from multiple cards dynamically;
- Difficult to intercept NFC communications from a distance;
- Ease of use for consumers in paying for goods;
- Simplifies access to back-end information; and
- Allows simple setup of new connections compared to other wireless protocols.

5. Limitations of NFC

Challenges of NFC technology include the following:
- Very short range of only a few inches precludes[1] many use cases;
- Slower than other protocols;
- Can limit usability for apps that require sensitive data on a smartphone;
- App innovation hindered by Apple and Google restrictions and tech implementations;
- Not suitable for location tracking; and
- Not as universal[2] and easy to integrate into venue ticketing apps as QR codes.

6. Differences Between NFC and Other Wireless Technologies

NFC complements a variety of other wireless technologies, including RFID, EMV (Europay, Mastercard and Visa), Bluetooth, UWB and QR codes. It also differs from these technologies.

NFC is best suited for authenticating transactions, unlocking doors and configuring other wireless connections. Other technologies work best in the following ways:
- RFID uses readers that can scan simple ID tags at long distances. Because it is unidirectional[3], this is best for reading toll tags, unlocking doors, authenticating passports or scanning inventory between more active readers and more passive tags.
- EMV allows for credit card transactions using a chip and an equipped payment terminal. While it also authenticates transactions, it is not as dynamic and interactive as NFC, which allows for contactless payments.
- Bluetooth offers a greater connection range than NFC but is less secure. It works best for connecting peripherals, such as headphones, to mobile devices and computers.
- UWB is a new technology and operates at a very low power using pulse patterns to keep from interfering with other wireless technologies. This allows it to send data quickly without losing accuracy. UWB excels in short-range location tracking, which works well for newer use cases, such as wireless car entry.

1　preclude [prɪˈkluːd] *vt*. 排除；阻止；妨碍
2　universal [ˌjuːnɪˈvɜːsl] *adj*. 普遍的，通用的
3　unidirectional [ˌjuːnɪdɪˈrekʃənəl] *adj*. 单向的

- QR codes need to be activated by the user by scanning an image with a smartphone's camera app, instead of a simple tap used by NFC. Businesses can easily generate QR codes for customers to retrieve[1] promotions, product manuals or webpages from printed tags. QR codes are less complex and are easier to include in emails without having to rely on integration with Apple or Google specific functionality.

As far as detection range, NFC can only identify whether a device is next to another. Bluetooth can recognize when an object is within a room. UWB can locate a remote controller buried between couch cushions to within 10 cm.

参考译文

Text A 无线网络如何工作

无线网络或无线局域网与有线网络的用途一样——把一组计算机连接起来。因为"无线"不需要昂贵的布线，所以其主要优点在于更容易、更快速和更廉价地组建网络。

相比而言，通过在办公室天花板和墙壁布线来组建网络更费人力，因此也花费更大。即使在某个地方已经布置了有线网络，无线网络也是扩展网络的有效方法。实际上，也不存在纯粹的无线网络，因为大多数无线网络都在某一点连回有线网络。

1. 基础

无线网络使用 RF（无线频率）技术，RF 是与无线电波传播相关的电磁频谱中的频率。当一个 RF 电流提供给天线时，一个电磁场就产生了，随后能通过空间传播。

无线网络的基础是一个叫作访问接入点（AP）的设备。访问接入点的主要任务是广播可以被计算机发现和"收听"到的无线信号。因为无线网络通常连到有线网络，访问接入点通常也起连接到有线网络的有效资源的作用，如连接到因特网。

要连接到访问接入点并与无线网络连接，计算机必须配置无线网络适配器。这些适配器通常内置于计算机中。如果没有内置，也可以在计算机的空扩展槽或 USB 接口插入适配器，或在笔记本计算机的 PC 卡槽加上它。这样几乎所有的计算机或笔记本都可以无线联网了。

2. 无线技术标准

因为无线网络有多个技术标准，因此在购买计算机之前一定要做足功课。最常见的无线技术标准包括以下几种：

- 802.11b：第一个广泛使用的无线组网技术，叫作 802.11b（更普遍地叫作 WiFi），已经出现了十余年了，但仍然在使用。
- 802.11g：在 2003 年，出现了叫作 802.11g 的后续版本，它提供了更好的性能（即速度和范围），依然是如今最常用的无线组网技术。
- 802.11n：它于 2009 年发布，提供了更快的无线传输速度和更好的覆盖范围。目前，802.11n 标准已经被更先进的无线标准取代，如 802.11ac（WiFi 5）和 802.11ax

---

1 retrieve [rɪˈtriːv] vt. 检索

（WiFi 6）。这些新标准提供更高的速度、更低的延迟和更好的性能。然而，由于 802.11n 标准的成本较低，并且许多设备仍然支持该标准，因此仍然有许多网络使用 802.11n 进行通信。对于一些低要求的应用场景，如家庭网络或小型企业网络，802.11n 仍然是一种可行的选择。

各种 WiFi（802.11b、802.11g 和 802.11n）都使用 2.4 GHz 的无线电频率，因此它们都被设计成相互兼容，所以可以在同一无线网络中使用不同标准的产品。如此一来，通常为了适应早期的设备要进行专门的配置,这也会相应地降低网络的整体性能。理想的情况是，全部无线网络设备、访问接入点和具有无线联网能力的全部计算机都使用相同的技术标准，并尽可能从一个销售商处购买。

3. 无线速度和范围

当购买一些无线网络硬件时，通常会提供基于所用无线组网标准的性能配置（传输数据的速率），还加上一些增强的技术。实际上，这些性能几乎总是被严重地夸大了。

802.11b、802.11g 和 802.11n 网络的官方速度分别是 11 兆位每秒、54 兆位每秒和 270 兆位每秒（Mbps），而这种速度在现实中简直无法实现。作为一般规律，可以认为在最理想的情况下，实现广告所宣称的三分之一性能。

还有一个值得重视的问题是无线网络被定义为共享网络，因此连入无线访问接入点的计算机越多，每个计算机能够收发的数据就越少。正如无线网络的速度可以有很多变化一样，其范围也可以改变。例如，802.11b 和 802.11g 正式的工作距离高达室内 100 米或者室外 300 米，但关键术语是"高达"。在现实中可能永远都接近不了这个数。

如所希望的，离访问接入点越近，信号就越强而且连接速度也更快。得到的无线网络范围和速度也取决于运行的环境，而这就带来干扰问题。

4. 无线干扰

在所有无线电通信中都存在干扰问题，无线网络也不例外。潜在的干扰在室内尤其严重，各种建筑材料（如混凝土、木头、纸面石膏板、金属、玻璃等）都可以吸收和反射无线电波，从而影响了无线网络信号的强度和连续性。同样地，像微波炉和无绳电话这样的设备也能产生干扰，因为它们运行在 2.4GHz 频段，与 802.11b/802.11g/802.11n 网络一样。不能完全避免干扰，但大多数情况下这不足以影响网络的可用性。如果有影响，通常可以通过重新部署无线组网硬件和使用特制天线使干扰最小化。

5. 无线网络的数据安全

有一台收音机才能接收当地无线电台，同样，要检测附近的无线网络也需要一台装备了无线设备的计算机。无法把网络对陌生人隐藏起来，但可以阻止未经授权的人连接到网络，并保护在网络中传输的数据不被偷窥。通过打开无线网络的加密功能，可以搅乱数据并控制对网络的访问。

无线网络硬件支持几种标准加密方案，但最常用的是 WEP、WPA 和 WPA2。WEP 最老并且安全性最差，应尽量不用。WPA 是一个相对更安全的选择，WPA2 被广泛视为最安全的协议。对于提高无线网络的安全性，使用 WPA2 是最佳选择。除非打算让公众访问你的无线网络——并把你的业务数据或个人数据置于危险中——那么你应该考虑强制使用加密术。

Unit 8

Text A

Top 5 IoT Development Platforms

1. ThingWorx 8 IoT Platform

ThingWorx 8 is a IoT development platform for developing, deploying, and managing IoT applications. It also supports various networking options and device management features.

The platform enables real-time data collection, secure storage, and analytics for visualization. It also offers a user-friendly development terrain and supports integration with enterprise systems. It ensures security through authentication and encryption mechanisms as well.

Pros:

- Rapid development: it offers many tools, libraries, and ready-made components that speed development. It also has a drag-and-drop feature that makes it easy to create IoT applications. You don't have to write a lot of code. This speeds up the process and enables IoT developers to launch applications in the market sooner.
- Scalability: it is designed to support a wide range of IoT installations. It can handle a large number of devices and data streams[1], whether it's for small test projects or large enterprise deployments.
- Connectivity: it can connect to networks using protocols like MQTT, HTTP and more. The device can easily integrate with different systems, sensors, and devices.
- Data analysis and visualization: it includes strong analytics capabilities to handle and analyze historical and real-time data from IoT devices.

---

1 Data streaming is the continuous transfer of data at a high rate of speed. Many data streams are collecting data from thousands of data sources at the same time. A data stream typically sends large clusters of smaller sized data records simultaneously.

- Security: it includes robust security features like access control, encryption, authentication, and authorization altogether.

Cons:
- Learning curve: while it provides various development tools, understanding the platform's concepts and capabilities may require some initial training.
- Cost: it is a platform that makes a profit. This is especially important for small businesses or startups that have limited money.
- Complications for basic use cases: it can handle large IoT projects, but it might be too complex for simple tasks like displaying data and basic networking.
- Customization restrictions: even though it has many ready-made parts, adjustments or specific needs might arise. The extent of the alteration will determine whether additional development effort or integration with external systems is necessary.

2. Microsoft Azure IoT Suite

Microsoft Azure IoT Suite is a platform on the cloud that helps connect and operate IoT devices. It also lets you analyze data and connect with AI services. It helps organizations securely connect and handle their IoT devices on a large scale.

The platform not only supports protocols but also offers device enrollment and configuration features. It allows ingestion and store housing of large volumes of IoT data, with real-time and batch analytics capabilities for extracting visualizations.

Azure IoT Suite integrates with Azure's machine literacy and AI services. It enables advanced analytics and prophetic capabilities. It also prioritizes security with the device-position security mechanisms and identity operation features.

Pros:
- Scalability: it handles big deployments easily. It can connect and manage millions of devices without any hassle. It also offers robust backend services, including data storage, analytics, and machine literacy capabilities, to support the growth of your IoT.
- Integration with Azure Services: this includes Azure Functions, Azure Stream Analytics, and Azure Machine Learning[1].
- Device management: it provides expansive device operation capabilities and allows you to cover, configure, and update IoT devices at scale. You also can set up, control firmware, and monitor device health.
- Security and compliance: it not only provides robust security features for your devices, data, and communications but also provides authentication, encryption, and part-grounded access control mechanisms to ensure secure data transmission and access.
- Analytics: it offers essential analytics capabilities to reuse and dissect the massive

---

1 Machine learning is a subfield of artificial intelligence that uses algorithms trained on data sets to create models that enable machines to perform tasks that would otherwise only be possible for humans, such as categorizing images, analyzing data, or predicting price fluctuations.

quantum of data IoT devices generate. You can work with Azure Stream Analytics and Machine Learning to gain precious visualizations, describe patterns, and make data-driven opinions.

Cons:
- Complexity: it is very complicated. It takes a lot of time to learn the platform.
- Cost: costs can increase, especially when dealing with many devices, data storage, and high-frequency data processing.
- Dependency on the cloud: this platform depends on the cloud, meaning the cloud services need an internet connection for IoT devices to communicate.
- Supplier lock-in: you rely on Microsoft's ecosystem and personal technologies when using Azure IoT Suite.

3. Google Cloud's IoT Platform

Google Cloud IoT Core is a comprehensive IoT App development platform that securely connects and manages devices. It supports various connectivity protocols, device management, and seamless integration with other Google Cloud services.

The platform ensures data ingestion and processing at scale, with strong security and identity management measures. It offers global scalability and has a different partner ecosystem.

Pros:
- Scalability: it has built an IoT platform that handles many IoT devices and data pipelines easily. Furthermore, it can scale up as needed, and ensure effective management of devices and data. It can also accommodate the growth of your IoT deployment as your requirements expand.
- Integration with Google Cloud services: it seamlessly integrates with other services handled by Google Cloud.
- Device operation: it offers robust device operation capabilities, including device enrollment, configuration, and over-the-air (OTA) firmware updates.
- Security: it not only provides erected-insecurity features to cover your IoT devices and data but also offers end-to-end encryption, authentication, and authorization mechanisms to ensure secure communication between devices and the cloud.
- Real-time data processing: it helps you quickly respond to events and make decisions based on real-time information from your IoT data. It also supports the ingestion and processing of data in real-time. This can be very important for time-sensitive operations and use cases.

Cons:
- Complexity: for newcomers, it can be complex.
- Cost: it has different pricing options, such as pay-as-you-go and dedicated plans.
- Learning curve: the platform is part of Google Cloud. Learning and using all the features and services may take some time.

- Supplier lock-in: by espousing Google Cloud's IoT platform, you become dependent on Google's structure and services.

4. IBM Watson IoT Platform

IBM Watson IoT App development platform is a cloud-based platform that facilitates the connection, operation, and analysis of IoT devices. It offers device connectivity, data operation, analytics, and visualization capabilities. It allows organizations to make decisions based on IoT data.

Pros:
- Scalability: it allows businesses to seamlessly connect and manage many IoT devices.
- Advanced analytics: it incorporates IBM Watson's cognitive computing capabilities, and enables advanced analytics and machine literacy.
- Security: it has robust security features that protect IoT devices and data.
- Integration with other IBM services: it integrates seamlessly with other IBM services.

Cons:
- Complexity: it requires specialized expertise to set up and configure the platform effectively.
- Cost: it can cost a lot for small businesses or startups with limited budgets.
- Limited device support: it can work with different devices and protocols. However, there may be some restrictions on device compatibility.

5. AWS IoT Platform

The AWS IoT platform by Amazon Web Services is an overall collection of services and tools for making and controlling IoT applications. It offers a safe and expandable framework for connecting devices and gathering data.

It also offers device management capabilities and a high-performing MQTT broker for effective messaging. Security features ensure data protection and device authentication.

Pros:
- Scalability: it offers a vastly scalable structure, which allows you to connect and manage billions of devices securely.
- Security: it incorporates robust security features, including end-to-end encryption, device authentication, and access control programs.
- Device management: it offers comprehensive device operation capabilities, enabling you to control and examine IoT devices.
- Integration with Other AWS services: it connects easily with other AWS services.
- Rule engine[1] and analytics: it provides a rule machine that allows you to define and

---

1 Rules engines or inference engines serve as pluggable software components which execute business rules that a business rules approach has externalized or separated from application code. This externalization or separation allows business users to modify the rules without the need for IT intervention.

execute conduct grounded on incoming device data.
- Device shadow: it supports device shadows, which are virtual representations of physical devices.

Cons:
- Complexity: it is complex to set up and configure.
- Cost: its services can increase costs, especially as the number of connected devices and data volume grows.
- Internet connectivity dependency: it depends on internet connectivity to communicate between devices and the cloud. However, if the network encounters issues or certain areas have limited internet connectivity, it can impact the performance of your IoT devices.
- Limited local processing: it offers cloud-based processing capabilities, but specific scripts might prefer processing data directly on edge.

New Words

develop	[dɪ'veləp]	vi. 开发
enable	[ɪ'neɪbl]	vt. 使能够，使可能
visualization	[ˌvɪʒʊəlaɪ'zeɪʃn]	n. 可视化，形象化
library	['laɪbrəri]	n. 库
drag-and-drop	[dræɡænddrɒp]	n.（鼠标的）拖放动作
create	[krɪ'eɪt]	vt. 建立，创造
code	[kəʊd]	n. 代码，密码
		vt. 编码
		vi. 为……编码
launch	[lɔːntʃ]	vt. 发布（应用程序）
		vi. 投入
authorization	[ˌɔːθəraɪ'zeɪʃn]	n. 授权，批准；批准（或授权）的证书
profit	['prɒfɪt]	n. 收益，利益；利润
startup	['stɑːtʌp]	n. 创业公司，新兴公司
complication	[ˌkɒmplɪ'keɪʃn]	n. 错杂，混乱
display	[dɪ'spleɪ]	v. 显示
		n. 显示器
customization	['kʌstəmaɪzeɪʃən]	n. 用户化，专用化，定制
restriction	[rɪ'strɪkʃn]	n. 限制，限定
adjustment	[ə'dʒʌstmənt]	n. 调节，调整
alteration	[ˌɔːltə'reɪʃn]	n. 变化，改变；变更
organization	[ˌɔːɡənaɪ'zeɪʃn]	n. 组织；机构；团体
		adj. 有组织的

enrollment	[ɪn'rəʊlmənt]	n. 注册；登记
ingestion	[ɪndʒestʃən]	n. 摄取
batch	[bætʃ]	n. 批处理，成批作业
extract	[ɪk'strækt]	v. 提取；选取；获得
prophetic	[prə'fetɪk]	adj. 预测的，预言的
prioritize	[praɪ'ɒrətaɪz]	vt. 按重要性排列，划分优先顺序；优先处理
expansive	[ɪk'spænsɪv]	adj. 易扩张的
firmware	['fɜːmweə]	n. 固件
essential	[ɪ'senʃl]	adj. 基本的；必要的；本质的
reuse	[ˌriː'juːz]	vt. 再用，重新使用
dissect	[dɪ'sekt]	vt. 仔细分析
massive	['mæsɪv]	adj. 大量的；大规模的
data-driven	['deɪtə'drɪvn]	adj. 数据驱动
opinion	[ə'pɪnjən]	n. 意见，主张；评价
comprehensive	[ˌkɒmprɪ'hensɪv]	adj. 广泛的；综合的
pipeline	['paɪplaɪn]	n. 管道
erected-in	[ɪ'rektɪd-ɪn]	adj. 内置的
event	[ɪ'vent]	n. 事件；活动
time-sensitive	[taɪm 'sensətɪv]	adj. 时间敏感性的
compatibility	[kəmˌpætə'bɪləti]	n. 互换性；通用性
expandable	[ɪk'spændəbl]	adj. 可扩展的，可扩大的
examine	[ɪg'zæmɪn]	vt. 检查
virtual	['vɜːtʃuəl]	adj. 虚拟的；实质上的，事实上的
representation	[ˌreprɪzen'teɪʃn]	n. 表现，表示，表现……的事物
dependency	[dɪ'pendənsi]	n. 依赖，依靠
encounter	[ɪn'kaʊntə]	v. 遭遇

Phrases

development platform	开发平台
ready-made component	现成的组件，制成的部件
a lot of	许多
data stream	数据流
test project	试验项目，测试项目
real-time data	实时数据
access control	访问控制
learning curve	学习曲线
initial training	初始训练，初步训练
device-position security mechanism	设备位置安全机制

backend service	后端服务
data storage	数据存储，数据保存
machine learning	机器学习
set up	建立；准备
deal with	处理
seamless integration	无缝集成
identity management	身份管理
device enrollment	设备登记
use case	用例
machine literacy	机器读写能力
gather data	收集数据
rule engine	规则引擎
local processing	本地处理

Abbreviations

OTA (Over-The-Air)	无线，空中下载技术

Analysis of Difficult Sentences

[1] It also has a drag-and-drop feature that makes it easy to create IoT applications.

本句中，that makes it easy to create IoT applications 是一个定语从句，修饰和限定 a drag-and-drop feature。在该从句中，that 是主语，指 a drag-and-drop feature；it 是形式宾语，真正的宾语是动词不定式短语 to create IoT applications；easy 作宾语补足语。

[2] IBM Watson IoT App development platform is a cloud-based platform that facilitates the connection, operation, and analysis of IoT devices.

本句中，that facilitates the connection, operation, and analysis of IoT devices 是一个定语从句，修饰和限定 a cloud-based platform。cloud-based 的意思是"基于云的"。

[3] It offers a vastly scalable structure, which allows you to connect and manage billions of devices securely.

本句中，which allows you to connect and manage billions of devices securely 是一个非限定性定语从句，对 a vastly scalable structure 进行补充说明。

[4] It provides a rule machine that allows you to define and execute conduct grounded on incoming device data.

本句中，that allows you to define and execute conduct grounded on incoming device data 是一个定语从句，修饰和限定 a rule machine。grounded on 的意思是"根据，基于"。

[5] It supports device shadows, which are virtual representations of physical devices.

本句中，which are virtual representations of physical devices 是一个非限定性定语从句，对 device shadows 进行补充说明。

Exercises

【EX.1】 Fill in the blanks according to the text.

1. ThingWorx 8 is a IoT development platform for _____, _____, and _____.
2. ThingWorx 8 offers many tools, libraries, and _____ that speed development. It also has a _____ that makes it easy to _____.
3. Azure IoT Suite integrates with _____ and _____. It enables advanced analytics and _____.
4. Azure IoT Suite not only provides _____ for your devices, data, and communications but also provides _____, _____, and part-grounded access control mechanisms to ensure _____ and access.
5. Google Cloud IoT Core ensures _____ and processing at scale, with strong security and _____. It offers _____ and has a different partner ecosystem.
6. Google Cloud IoT Core helps you quickly respond to _____ and make decisions based on _____ from your IoT data. It also supports the _____ and processing of data in real-time.
7. IBM Watson IoT App development platform is a _____ platform that facilitates the connection, _____, and _____ of IoT devices. It offers device connectivity, _____, analytics, and visualization capabilities.
8. IBM Watson IoT Platform requires _____ to set up and _____ the platform effectively.
9. AWS IoT platform offers a safe and expandable _____ for connecting devices and _____. It also offers _____ capabilities and a high-performing MQTT broker for effective messaging.
10. AWS IoT platform incorporates robust security features, including _____, _____, and _____. It supports device shadows, which are _____.

【EX.2】 Translate the following terms or phrases from English into Chinese and vice versa.

1.	customization	1.	
2.	developer	2.	
3.	enable	3.	
4.	expandable	4.	
5.	extract	5.	
6.	firmware	6.	
7.	massive	7.	
8.	access control	8.	
9.	data stream	9.	
10.	adjustment	10.	
11.	开发平台	11.	

12.	机器学习	12.	_____
13.	用例	13.	_____
14.	*adj*. 虚拟的；实质上的，事实上的	14.	_____
15.	*n*. 互换性；通用性	15.	_____

【EX.3】 Translate the following sentences into Chinese.

1. You should create a new directory and put all your files into it.
2. As we all see, visualization methods are varied.
3. You can move address lists to create a new hierarchy, using a drag-and-drop operation.
4. The bar codes on the products are read by lasers.
5. An error code will be displayed if any invalid information has been entered.
6. I've made a few adjustments to the design.
7. I've simply extracted a few figures.
8. All the products are labelled with comprehensive instructions.
9. Firmware update includes the latest fixes and new features.
10. Data-driven based methods are important branches in the field of process monitoring.

【EX.4】 Complete the following passage with appropriate words or expressions in the box.

configurations	compatible	ensure	spectrum	plug
broadcasting	regulation	modem	marketing	center

"WiFi" stands for "wireless fidelity". This is a __1__ name and bares no relation particularly to the technology. WiFi networks use the 2.4 GHz radio __2__ to transmit data. For comparison's sake your mobile phone operates at around 800 Mhz and FM radio operates at up to 100 Mhz.

At the __3__ of a WiFi network is the WiFi router. Although routers come in many __4__, one of the most flexible arrangements is an adsl __5__, Ethernet router and WiFi base station all rolled into one device.

This gives you the fall back position of Ethernet or the ability to __6__ in a non WiFi network printer. For security reasons you should __7__ your WiFi router is capable of NAT, WPA and the ability to turn off SSID __8__.

There are a number of standards in WiFi; a, b & g. g is the current standard. Many g devices are backwardly __9__ with b devices. WiFi is a global standard and now there are many manufacturers of WiFi products; thus the price has come down in recent years.

Different countries have different supervision systems concerning WiFi. There are more __10__ in Australia than in the United States.

【EX.5】 Translate the following passage into Chinese.

What Is WiFi?

WiFi (short for "wireless fidelity") is a term for certain types of wireless local area network (WLAN) that use specifications in the 802.11 family. The term WiFi was created by an organization called the WiFi Alliance, which oversees tests that certify product interoperability. A product that passes the alliance tests is given the label "WiFi certified" (a registered trademark).

Originally, WiFi certification was applicable only to products using the 802.11b standard. Today, WiFi can apply to products that use any 802.11 standard. The 802.11 specifications are part of an evolving set of wireless network standards known as the 802.11 family. WiFi has gained acceptance in many businesses, agencies, schools, and homes as an alternative to a wired LAN. Many airports, hotels, and fast-food facilities offer public access to WiFi networks. These locations are known as hot spots. Many charge a daily or hourly rate for access, but some are free. An interconnected area of hot spots and network access points is known as a hot zone.

Unless adequately protected, a WiFi network can be susceptible to access by unauthorized users who use the access as a free Internet connection. Any entity that has a wireless LAN should use security safeguards such as the wired equivalent privacy (WEP) encryption standard, the more recent WiFi protected access (WPA), internet protocol security (IPsec), or a virtual private network (VPN).

Text B

Programming Languages for IoT

1. Java

Java is the most well-known and popular language among developers and it is probably the best choice for IoT developers as a programming language for its write once run anywhere (WORA) principle. Developers can use it to create and debug code on their computer, then transfer the code to any chip through a Java virtual machine (JVM). So, the code written in Java can be run on any place where JVMs are used and on any other minor/smaller machines as well. Today Java SE Embedded is more focused on embedded development. It is one of the best choices for IoT applications. Java's object-oriented and least hardware dependency and its' hardware support libraries have made it one of the best choices for IoT development.

Some of the benefits of using Java for IoT include:

- Platform independence[1]: Java code can run on any device that has a Java virtual machine

---

1 Platform independent refers to software that runs on a variety of operating systems or hardware platforms. It is the opposite of platform dependent, which refers to software that is only to run on one operating system or hardware platform.

(JVM)[1] installed, making it easy to develop and deploy IoT applications on a wide range of platforms and devices.
- Security: Java has built-in security features, such as automatic memory management and type checking, that help prevent common security vulnerabilities and ensure the safety of IoT devices and networks.
- Scalability: Java is a platform-independent language that can be used to build large-scale IoT systems, making it a good choice for IoT projects that require high performance and scalability.
- Large community: Java has a large and active community, which provides a wealth of resources, including libraries, tutorials, and support forums. This makes it easier for developers to find the resources they need to build IoT applications.

In IoT development, Java is often used to develop software for gateways and other intermediate devices, as well as to create applications that collect and process data from IoT devices. Java is also used to develop applications that control and monitor IoT devices, as well as to implement security protocols for IoT networks.

2. C

C language which is known as the mother of every programming language is still the most essential programming language for IoT development. It is a popular choice for developing software for IoT devices due to its low-level nature and efficient use of system resources. Some of the benefits of using C for IoT include:
- Efficient use of system resources: C is a low-level language, which means it is closer to machine code and operates more efficiently than high-level languages[2]. This makes it a good choice for IoT devices with limited processing power and memory.
- Portability: C code can be easily ported to different platforms and devices, making it a versatile choice for IoT development.
- Legacy code: C has a long history and is widely used, making it easy for developers to find existing code and libraries that can be reused in IoT projects.
- Performance: C is a fast language, making it well-suited for real-time systems, such as those used in IoT.

In IoT development, C is often used to program microcontrollers and other low-level hardware components, as well as to write firmware for IoT devices. It is also used to develop software for gateways and other intermediate devices that act as intermediaries between IoT devices and the cloud or other higher-level systems.

---

1 The Java virtual machine, or JVM, is a key component of the Java programming language, providing a platform-independent environment for the execution of Java code across most major hardware, operating systems, and software architectures.

2 A high-level language (HLL) is a programming language such as C, FORTRAN, or Pascal that enables a programmer to write programs that are more or less independent of a particular type of computer. Such languages are considered high-level because they are closer to human languages and further from machine languages.

3. Python

Python is the perfect language for data-intensive applications. For example, if there is a lot of data processing work, Python is the best to use. It is the best choice for IoT development which includes data application, data science, and analytics capabilities to the edge. Now it is the language of choice for one of the most popular microcontrollers on the market. Python is quite a simple, flexible, and uncomplicated language. Its large set of libraries and tools have made it compatible with IoT.

Some of the benefits of using Python for IoT include:
- Ease of use: Python has a simple and intuitive syntax, making it easy for developers to write and understand code, especially for those with limited programming experience.
- Large community: Python has a large and active community, which provides a wealth of resources, including libraries, tutorials, and support forums. This makes it easier for developers to find the resources they need to build IoT applications.
- Rich libraries: Python has a vast collection of libraries that support a wide range of tasks, from data processing and visualization to machine learning and network communication. This makes it easier for developers to build sophisticated IoT applications.
- Cross-platform compatibility: Python is a cross-platform language that runs on many different operating systems, including Windows, Linux, and macOS, making it easy to develop and deploy IoT applications on a variety of platforms.

In IoT development, Python is often used to develop software for gateways and other intermediate devices, as well as to create applications that communicate with and control IoT devices. Python is also used to develop applications that collect and process data from IoT devices and perform data analysis and visualization.

4. JavaScript

JavaScript has become one of the best choices for IoT developers. When much focus is on the gateway and edge nodes as well as on the IoT cloud applications, JavaScript is a popular choice for developers. JavaScript is a high-level, dynamic programming language that is widely used for IoT development, particularly for Web-based IoT applications. Some of the benefits of using JavaScript for IoT include:
- Easy to learn: JavaScript is a simple and intuitive language that is easy to learn, making it a good choice for developers of all skill levels.
- Dynamic nature: JavaScript is a dynamic language, which makes it easy to change and modify code on the fly, without the need for recompilation. This makes it a good choice for IoT applications that require real-time updates.
- Cross-platform compatibility: JavaScript can run in any Web browser, making it easy to develop and deploy IoT applications on a wide range of platforms and devices.
- Rich libraries: JavaScript has a rich collection of libraries, including popular frameworks such as Node.js and React, that support a wide range of tasks, from web development to

machine learning. This makes it easier for developers to build sophisticated IoT applications.

In IoT development, JavaScript is often used to develop Web-based applications that control and monitor IoT devices, as well as to create user interfaces for IoT systems. It is also used to develop server-side applications that collect and process data from IoT devices and perform data analysis and visualization.

5. Swift

Swift is a modern, high-performance programming language developed by Apple Inc. It is widely used for developing software for iOS and macOS, and is also used in some IoT applications. Some of the benefits of using Swift for IoT include:

- High performance: Swift is designed to be fast and efficient, making it a good choice for IoT applications that require high performance.
- Safe and secure: Swift has built-in safety features, such as type checking and automatic memory management, that help prevent common security vulnerabilities and ensure the safety of IoT devices and networks.
- Easy to learn: Swift is a simple and intuitive language that is easy to learn, making it a good choice for developers of all skill levels.
- Integration with Apple ecosystems: Swift is closely integrated with the iOS and macOS ecosystems, making it easy to develop and deploy IoT applications on Apple devices.

In IoT development, Swift is often used to develop software for Apple-based IoT devices, such as the Apple Watch and the HomePod. Swift is also used to develop applications that control and monitor IoT devices, as well as to implement security protocols for IoT networks.

The choice of programming language completely depends on the developers and requirements. All of the top programming languages have their benefits and use cases. But it is widely accepted that for devices, people choose C and C++; for gateway and IoT applications, Java and Python; for cloud, Java and JavaScript are the best choices.

New Words

developer	[dɪ'veləpə]	n. 开发者
object-oriented	['ɒbdʒekt 'ɔːrɪəntɪd]	adj. 面向对象的
large-scale	[lɑːdʒ skeɪl]	adj. 大规模的，大范围的
forum	['fɔːrəm]	n. 论坛，讨论会
low-level	[ləʊ'levl]	adj. 低级的
efficient	[ɪ'fɪʃnt]	adj. 有效率的；（直接）生效的
versatile	['vɜːsətaɪl]	adj. 多用途的；多功能的
intermediary	[ˌɪntə'miːdɪəri]	n. 中间件
uncomplicated	[ʌn'kɒmplɪkeɪtɪd]	adj. 简单的，不复杂的
cross-platform	[krɒs 'plætfɔːm]	n. 跨平台

dynamic	[daɪ'næmɪk]	*adj.* 动态的
recompilation	[rekəmpɪ'leɪʃn]	*n.* 重新编译，再编译
browser	['braʊzə]	*n.* 浏览器
server-side	['sɜːvə saɪd]	*adj.* 服务器端的
high-performance	[haɪpə'fɔːməns]	*adj.* 高性能的

Phrases

programming language	程序设计语言，编程语言
embedded development	嵌入式开发
support library	支持库
platform independence	平台独立性
memory management	内存管理
platform-independent language	平台独立语言
intermediate device	中间设备
system resource	系统资源
low-level language	低级语言
high-level language	高级语言
ported to	移植到
real-time system	实时系统
data-intensive application	数据密集型应用程序
data processing	数据处理
intuitive syntax	直观的语法
network communication	网络通信
a variety of	多种的
dynamic language	动态语言
on the fly	（计算机）运行中
real-time update	实时更新
cross-platform compatibility	跨平台兼容性
web development	网络开发
web-based application	基于网络的应用程序
be designed to	被设计用于

Abbreviations

| WORA (Write Once Run Anywhere) | 一次写成，随处可用 |
| JVM (Java Virtual Machine) | Java 虚拟机 |

Exercises

【EX.6】 Answer the following questions according to the text.

1. What is Java?

2. What are some of the benefits of using Java for IoT?
3. Why is C a popular choice for developing software for IoT devices?
4. What are some of the benefits of using C for IoT?
5. What are some of the benefits of using Python for IoT?
6. What is Python often used to do in IoT development?
7. What are some of the benefits of using JavaScript for IoT?
8. What is JavaScript often used to do in IoT development?
9. What is Swift?
10. What is Swift often used to do in IoT development?

【EX.7】 Translate the following terms or phrases from English into Chinese and vice versa.

1. cross-platform
2. intermediary
3. object-oriented
4. recompilation
5. low-level
6. dynamic
7. data processing
8. embedded development
9. high-level language
10. intermediate device
11. 内存管理
12. 网络通信
13. 实时系统
14. 系统资源
15. 平台独立性

【EX.8】 Translate the following sentences into Chinese.

1. Each software developer should study and master some modern software development processes.
2. Developers use this environment to perform integration tests among all system components.
3. The design of this system adopted compound structure and object-oriented thought mainly.
4. Network management system is a large-scale and complex system software engineering.
5. Network forums emerged quickly with the rapid rise of the internet.
6. It allows for parallel and independent development of client-side and server-side.
7. This cross-platform approach raises many difficulties, a major one being file access.
8. This protects against stack, buffer, and function pointer overflows, all without recompilation.
9. In your Web browser's cache are the most recent Web files that you have downloaded.

10. Press the "reload" button on your web browser to refresh the site and get the most current version.

Reading Material

Top 10 IoT Tools

IoT has changed the way we interact with technology by enabling seamless connection and communication between devices. As IoT continues to evolve, developers and organizations have access to a wide range of tools that provide efficient development and deployment of IoT solutions.

Various tools, including hardware and software platforms, network analyzers and IoT-specific platforms, empower developers to build, connect, analyze and monitor IoT solutions efficiently. These tools accelerate[1] development, ensure data security, and optimize IoT application performance.

1. Features of IoT Tools

- Device management: registering, configuring and monitoring IoT devices, including provisioning, firmware updates, and remote management.
- Data visualization: customizable[2] dashboards[3] for real-time monitoring and analysis of IoT data, facilitating data-driven[4] decisions.
- Connectivity and integration: support for various IoT protocols, cloud platforms and databases, enabling seamless communication between devices and services.
- Rule engine and automation: definition of rules, triggering actions and task automation based on specific conditions or events in the IoT system.
- Security and privacy: authentication, access control, encryption and secure communication protocols to safeguard IoT devices and data.
- •Scalability and performance: optimization for large-scale IoT deployments, managing numerous devices and processing high volumes of data.
- •Analytics and machine learning: advanced capabilities for data processing, anomaly detection, predictive modeling and intelligent insights.

2. IoT Development Tools

IoT revolutionizes industries and daily life by connecting devices and enabling data-driven decision-making. Developers and organizations have a wide range of IoT tools for efficient development and deployment.

2.1 Arduino

Arduino produces electronic devices and software for the IoT market, offering top-notch[5]

---

1 accelerate [əkˈseləreɪt] vt. 加速，（使）加快，（使）增速
2 customizable [ˈkʌstəmaɪzəbəl] adj. 可定制的
3 dashboard [ˈdæʃbɔːd] n. 仪表板，仪表盘
4 data-driven [ˈdeɪtəˈdrɪvn] adj. 数据驱动的
5 top-notch [tɒp nɒtʃ] adj. 拔尖的，一流的，一等的

hardware for a wide range of projects. It is a leading company in IoT tools, allowing easy construction of functional and innovative robotics and home automation projects.

Overview and features:
- Arduino boards: Arduino offers diverse microcontroller boards for IoT projects with varying specifications.
- Arduino IDE: an integrated development environment (IDE) offers a user-friendly interface that facilitates the process of writing and uploading code to Arduino boards.
- Libraries and examples: Arduino offers an extensive collection of pre-existing[1] code and samples, streamlining the creation of IoT applications.
- Community support: the Arduino community consists of a vibrant[2] group of developers who actively engage with one another, exchange knowledge and provide assistance to fellow users.

2.2 Apache NetBeans

Apache NetBeans is a versatile[3] IDE that is open-source and offers support for various programming languages. It offers a strong platform for the development of IoT applications.

Overview and features:
- Language support: NetBeans supports multiple programming languages like Java, JavaScript, C++ and more, empowering developers to choose the ideal language for their IoT projects.
- Project management: NetBeans streamlines IoT app development with templates, code completion and debugging.
- Plugins and extensions: NetBeans has a thriving plugin ecosystem for IoT-specific requirements.
- Collaboration tools: NetBeans facilitates team collaboration on IoT projects with version control, code sharing and collaboration features.

2.3 Kinoma

Kinoma is an IoT toolset that eases the development of embedded apps and devices. It includes a JavaScript framework and hardware kits for fast prototyping[4].

Overview and features:
- Kinoma Create: Kinoma Create is a hardware development kit with a programmable device and touch-enabled color display for fast IoT prototype building.
- JavaScript framework: Kinoma's JavaScript framework enables developers to create IoT applications and user interfaces in a familiar and accessible programming environment.
- Integrated tools: Kinoma Studio simplifies the creation and management of IoT applications

---

1　pre-existing [ˌpriːɪɡˈzɪstɪŋ] adj. 先已存在的，早已存在的
2　vibrant [ˈvaɪbrənt] adj. 充满生机的，生气勃勃的
3　versatile [ˈvɜːsətaɪl] adj. 多用途的，多功能的
4　fast prototyping：快速原型设计

with a visual interface for building and connecting workflows[1].
- Internet connectivity: Kinoma enables IoT devices to connect and communicate with other devices, cloud services and web APIs using various communication protocols.

2.4 MQTT (Message Queueing Telemetry Transport)

MQTT is a lightweight messaging protocol for efficient IoT communication, enabling real-time data exchange with low bandwidth consumption.

Overview and features:
- Lightweight: MQTT is optimized for resource-constrained IoT devices, with low processing power and limited bandwidth.
- Publish-subscribe model[2]: MQTT uses a publish-subscribe model, where devices publish messages to topics and others subscribe to receive them.
- QoS[3] levels: MQTT has three levels of QoS for reliable message delivery based on application requirements.
- Retained messages: MQTT retains the latest value, ensuring subscribers receive the most recent message even if they were offline during publishing.

2.5 Wireshark

Wireshark, a powerful network protocol analyzer, is indispensable[4] for developers to monitor and analyze network traffic. It provides detailed insights into the communication between IoT devices and networks, making it essential for IoT application development and debugging.

Overview and features:
- Network traffic analysis: Wireshark captures and analyzes network packets, revealing IoT device and network communication.
- Protocol support: Wireshark analyzes IoT network traffic with support for protocols like MQTT, HTTP, CoAP, ZigBee, and more.
- Filtering and search: Wireshark simplifies the analysis by allowing users to filter[5] and search for specific packets or data patterns.
- Packet decoding: Wireshark decodes network packets into a readable format, helping to identify issues and understand information flow.

2.6 Mainspring

Mainspring is an IoT development platform that operates on the cloud, making the creation and management of IoT applications easier. It offers a wide range of tools and services that streamline the development and deployment of IoT projects.

---

1 workflow ['wɜːkfləʊ] *n.* 工作流程
2 publish-subscribe model：发布-订阅模型
3 QoS (Quality of Service)：服务质量
4 indispensable [ˌɪndɪ'spensəbl] *adj.* 不可缺少的，绝对必要的
5 filter ['fɪltə] *v.* 过滤

Overview and features:
- Rapid application development: mainspring enables coding-free IoT application design with a user-friendly visual interface and drag-and-drop[1] functionality.
- Device management: users can remotely monitor, control and update IoT devices using the platform's robust device management capabilities.
- Data analytics: with its built-in data analytics features, Mainspring empowers users to extract valuable insights from IoT data and make data-driven decisions.
- Security and scalability: the platform places emphasis on ensuring the security of data and provides scalability choices to accommodate the expansion of IoT deployments.

2.7 Node-RED

Node-RED is a tool for visual programming and connecting IoT hardware devices, APIs and online services. It offers a flow editor in a Web browser, enabling users to create IoT applications through the visual connection of nodes.

Overview and features:
- Flow-based programming: Node-RED uses flow-based programming to connect nodes and create IoT applications.
- Extensive library of nodes: Node-RED has a wide selection of pre-built nodes for easy integration with IoT devices, protocols and services.
- Real-time monitoring: the platform offers a real-time dashboard to monitor and visualize IoT data.
- Easy integration: Node-RED integrates seamlessly with various tools and platforms, making it versatile for IoT development.

2.8 Eclipse IoT

Eclipse IoT is an open-source platform that simplifies IoT application development and management with various tools and frameworks.

Overview and features:
- Eclipse IoT projects: Eclipse IoT provides projects and frameworks for IoT development, covering device connectivity, data management and security.
- Interoperability[2]: the platform emphasizes interoperability, enabling developers to integrate IoT devices, protocols and cloud services from different vendors.
- Scalability and performance: Eclipse IoT offers scalable tools and frameworks for the efficient management of large-scale IoT deployments.
- Community support: Eclipse IoT has a vibrant community of developers who contribute to the projects, share knowledge and offer support.

2.9 ThingsBoard IoT

ThingsBoard IoT is an open-source platform for collecting, analyzing and visualizing data

---

1　drag-and-drop [dræg ænd drɔp] n.（鼠标的）拖放动作
2　interoperability [ˈɪntərɒpərəˈbɪləti] n. 互用性

from IoT devices, providing powerful features for managing and monitoring IoT deployments.

Overview and features:
- Device management: ThingsBoard IoT manages IoT devices with registration[1], configuration and monitoring.
- Data visualization[2]: customize dashboards and widgets for real-time visualization of IoT data, enabling data-driven insights.
- Rule engine: define rules and trigger actions based on conditions or events using ThingsBoard's rule engine.
- Integration options: flexibly integrate with IoT protocols, cloud platforms and databases for building IoT applications.

2.10　Zetta IoT

Zetta IoT is an open-source platform for scalable IoT systems, treating everything as an API for unified device and service interaction.

Overview and features:
- API-centric approach: Zetta IoT treats every IoT device and service as an API for seamless integration and interaction.
- Scalability and flexibility: manage a large number of devices in a distributed IoT system with Zetta IoT's scalable design.
- Real-time streaming: Zetta IoT enables real-time streaming of IoT data for instant processing and analysis.
- Web-based interface: access and manage IoT devices through a Web-based interface on any browser[3].

3. Conclusion

These top 10 IoT tools support diverse functionalities for developing and managing IoT applications. Having tools for a specific domain is a great thing. These tools can enhance your workflow and are easy to learn.

参考译文

Text A　五大物联网开发平台

1. ThingWorx 8 物联网平台

ThingWorx 8 是一个用于开发、部署和管理物联网应用程序的物联网开发平台。它还支持各种网络选项和设备管理功能。

该平台支持实时数据收集、安全存储和可视化分析。它还提供了用户友好的开发平台，

1　registration [ˌredʒɪˈstreɪʃn] n. 登记，注册
2　data visualization：数据可视化
3　browser [ˈbraʊzə] n. 浏览器

并支持与企业系统的集成。它还通过身份验证和加密机制确保安全性。

优点：
- 快速开发：它提供了许多可以加速开发的工具、库和现成的组件。它还具有拖放功能，可以轻松创建物联网应用程序，而不必编写大量代码。这加快了应用程序的开发过程，使物联网开发人员能够更快地将其推到市场上。
- 可扩展性：它旨在支持广泛的物联网安装。它可以管理大量的设备和处理大量的数据流，无论是小型测试项目还是大型企业部署都可应用。
- 连接性：它可以使用 MQTT、HTTP 等协议连接到网络。该设备可以轻松地与不同的系统、传感器和设备集成。
- 数据分析和可视化：它具有强大的分析功能，可处理和分析来自物联网设备的历史和实时数据。
- 安全性：它包括强大的安全功能，如访问控制、加密、身份验证和授权。

缺点：
- 学习曲线：虽然它提供了各种开发工具，但了解该平台的概念和功能可能需要一些初步培训。
- 成本：这是一个盈利平台，这对于资金有限的小型企业或初创企业尤其重要。
- 基本用例的复杂性：它可以管理大型物联网项目，但对于显示数据和基本网络等简单任务来说可能过于复杂。
- 定制限制：即使有许多现成的部件，也可能会出现调整或特定需求。更改的程度将决定是否需要额外的开发工作或与外部系统的集成。

2. Microsoft Azure IoT Suite

Microsoft Azure IoT Suite 是一个帮助连接和操作物联网设备的云平台。它还可以让你分析数据并与人工智能服务连接。它可以帮助组织大规模安全地连接和操作其物联网设备。

该平台不仅支持协议，还提供设备注册和配置功能。它允许摄取和存储大量物联网数据，并具有用于提取可视化数据的实时和批量分析功能。

Azure IoT Suite 与 Azure 的机器读写及人工智能服务集成。它支持高级分析和预测功能。它还优先考虑设备位置安全机制和身份操作功能的安全性。

优点：
- 可扩展性：它可以轻松处理大型部署。它可以轻松地连接和管理数百万台设备。它还提供强大的后端服务，包括数据存储、分析和机器读写能力，以支持物联网的增长。
- 与 Azure Services 集成：这包括 Azure Functions、Azure Stream Analytics 和 Azure Machine Learning。
- 设备管理：提供丰富的设备操作能力，能够大规模覆盖、配置和更新物联网设备，还可以设置、控制固件和监控设备运行状况。
- 安全性和合规性：它不仅为设备、数据和通信提供了强大的安全功能，还提供了身份验证、加密和部分访问控制机制，以确保数据传输和访问的安全。
- 分析：它提供必要的分析功能来重用和剖析物联网设备生成的大量数据。可以使用

Azure Stream Analytics 和 Azure Machine Learning 来获得宝贵的可视化效果、描述模式并提出数据驱动的意见。

缺点：
- 复杂性：它非常复杂，需要花很多时间来学习这个平台。
- 成本：成本可能会增加，尤其是在处理许多设备、数据存储和高频数据处理时。
- 对云的依赖：该平台依赖云，这意味着云服务需要互联网连接才能使物联网设备进行通信。
- 供应商锁定：使用 Azure IoT Suite 时，依赖 Microsoft 的生态系统和个人技术。

3. Google Cloud 的物联网平台

Google Cloud IoT Core 是一个全面的物联网应用开发平台，可安全连接和管理设备。它支持各种连接协议、设备管理并与其他 Google Cloud 服务无缝集成。

该平台通过强大的安全和身份管理措施确保大规模的数据摄取和处理。它提供全球可扩展性，并拥有不同的合作伙伴生态系统。

优点：
- 可扩展性：它构建了一个物联网平台，可以轻松管理许多物联网设备和数据管道。此外，它可以根据需要进行扩展，并确保有效地管理设备和数据。随着需求的扩展，它还可以适应物联网部署的增长。
- 与 Google Cloud 服务集成：它与 Google Cloud 处理的其他服务无缝集成。
- 设备操作：它提供强大的设备操作功能，包括设备注册、配置和无线（OTA）固件更新。
- 安全性：它不仅提供内置的安全功能来保护物联网设备和数据，还提供端到端加密、身份验证和授权机制，以确保设备和云之间的安全通信。
- 实时数据处理：它可以帮助快速响应事件并根据物联网数据的实时信息做出决策。它还支持实时数据的摄取和处理，这对于时间敏感的操作和用例非常重要。

缺点：
- 复杂性：对于新手来说，可能会复杂。
- 成本：有不同的定价选项，如即用即付和专用计划。
- 学习曲线：该平台是 Google Cloud 的一部分，学习和使用所有功能和服务可能需要一些时间。
- 供应商锁定：通过支持 Google Cloud 的物联网平台，会变得依赖 Google 的结构和服务。

4. IBM Watson 物联网平台

IBM Watson 物联网应用开发平台是一个基于云的平台，有助于物联网设备的连接、操作和分析。它提供设备连接、数据操作、分析和可视化功能。它允许组织根据物联网数据做出决策。

优点：
- 可扩展性：它允许企业无缝连接和管理许多物联网设备。

- 高级分析：它结合了 IBM Watson 的认知计算功能，并支持高级分析和机器读写。
- 安全性：它具有强大的安全功能，可以保护物联网设备和数据。
- 与其他 IBM 服务集成：它与其他 IBM 服务无缝集成。

缺点：
- 复杂性：需要专业知识才能有效地设置和配置平台。
- 成本：对于预算有限的小型企业或初创公司来说，成本可能会很高。
- 有限的设备支持：它可以与不同的设备和协议配合使用。但是，设备兼容性可能存在一些限制。

5. AWS 物联网平台

Amazon Web Services 的 AWS 物联网平台是用于创建和控制物联网应用的服务和工具的整体集合。它为连接设备和收集数据提供了一个安全且可扩展的框架。

它还提供设备管理功能和高性能 MQTT 代理，以实现有效的消息传递。安全功能可切实保护数据和进行设备身份验证。

优点：
- 可扩展性：它提供了高度可扩展的结构，使可以安全地连接和管理数十亿台设备。
- 安全性：它包含强大的安全功能，包括端到端加密、设备身份验证和访问控制程序。
- 设备管理：它提供全面的设备操作能力，使能够控制和检查物联网设备。
- 与其他 AWS 服务集成：它可以轻松地与其他 AWS 服务连接。
- 规则引擎和分析：它提供了一个规则机，允许根据传入的设备数据来定义行为并执行。
- 设备影子：它支持设备影子，即物理设备的虚拟表示。

缺点：
- 复杂性：设置和配置都很复杂。
- 成本：其服务可能会增加成本，尤其是当连接设备数量和数据量增长时。
- 互联网连接依赖性：设备和云之间的通信依赖互联网连接。但是，如果网络遇到问题或某些区域的互联网连接有限，则可能会影响物联网设备的性能。
- 本地处理能力有限：尽管它提供基于云的处理功能，但特定脚本可能更喜欢直接在边缘处理数据。

Unit 9

Text A

IoT Security

1. What Is IoT Security?

IoT security is an umbrella term for the tools and strategies that protect devices connected to the cloud and the network they use to connect to each other. Its main goals are to keep user data safe, stop cyber attacks and keep devices running smoothly.

2. Importance of IoT Security

The significance of IoT security cannot be overstated due to various factors. IoT devices often store critical information, such as financial and personal data, which must be safeguarded. Any breach in security may expose this data, leading to detrimental consequences like identity theft and financial harm.

IoT devices are crucial to essential infrastructure, such as power grids, transportation systems, and healthcare.Any unauthorized access[1] to these systems can result in severe repercussions, such as power failures, disruptions in transportation, and potential loss of lives.

3. IoT Security Threats

IoT security is subjected to a wide range of risks that are continually changing. Here are a few prevalent IoT security threats that demand our attention:
- Botnets: botnets are networks of compromised devices that cyber criminals can controll to launch distributed denial of service (DDoS) attacks, steal data, or engage in other malicious activities. IoT devices are often used in botnets due to their large numbers and

---

1 Unauthorized access is when a person who does not have permission to connect to or use a system gains entry in a manner unintended by the system owner. The popular term for this is "hacking".

weak security.
- Malware: malware is software that is designed to infiltrate and damage computers and other devices. IoT devices are often vulnerable to malware attacks, which can compromise their functionality and steal data.
- Physical tampering: IoT devices can be physically tampered with to gain unauthorized access to the device or network. This can involve breaking into the device itself or intercepting signals between the device and the network.
- Data breaches: IoT devices frequently retain confidential information, including personal data, financial records, and other sensitive details. If there is a breach in security, the exposure of this data can result in severe consequences, such as identity theft, financial hardships, and other adverse outcomes.
- Weak passwords: weak passwords are a common IoT security threat, as many IoT devices use default passwords that are easy to guess or crack. This makes it easy for attackers to gain access to IoT devices and networks.

4. Types of IoT Security

There are three distinct types of IoT security:
- •Network security: network security is any activity designed to protect the usability and integrity of your network and data. It includes both hardware and software security. Its target is to deal with a variety of threats. It stops others from entering or spreading on your network. Effective network security manages access to the network.
- Device security: device security is the safeguarding of internet-connected devices, such as mobile phones, laptops, PCs, tablets, IoT devices, from cyber threats and unauthorized access. Device security typically requires strong authentication, mobile device management tools, and restricted network access.
- Firmware security: firmware security begins by conducting an in-depth examination of an IoT device's firmware to detect potential vulnerabilities within its code. Firmware security uses instrumentation to monitor the state of binaries on the device as they run. This deeper visibility enables firmware security solutions to identify attempted attacks and remediate the damaged firmware based on anomalous behavior by the monitored binary and to identify and prevent even novel IoT vulnerabilities.

5. Best Practices for IoT Security

5.1 Device Authentication and Access Control

Ensuring the security of IoT involves two essential elements, namely device authentication and access control. Device authentication validates the identity of IoT devices, which permits network access solely to authorized devices. Access control, in turn, governs the extent of privileges granted to individual devices or users. To enhance security, IoT devices should be configured to employ robust authentication methods, such as biometric authentication, two-factor authentication[1], or digital certificates[2], for verifying the identity of both devices and users.

Access control should also be enforced to limit the access privileges of each device or user based on their role and level of trust. This can be achieved by using access control lists (ACLs), which specify which devices or users are authorized to access specific resources or perform certain actions. Additionally, IoT devices should be configured to log all access attempts and activities to enable audit trails and traceability.

5.2 Encryption

Encryption is a crucial component of IoT security that ensures the confidentiality and integrity of the data transmitted between IoT devices and networks. Encryption entails using an encryption algorithm and a secret key to convert plain text data into ciphertext. Only authorized users with the correct key can decrypt the ciphertext and access the plain text data.

IoT devices should use strong encryption algorithms such as advanced encryption standard (AES) or secure hash algorithm (SHA) to secure data in transit and at rest. Additionally, IoT devices should be configured to use secure communication protocols such as transport layer security (TLS) or secure sockets layer (SSL) to protect data that is transmitted over the internet. Encryption keys should be securely stored and managed to prevent unauthorized access and misuse.

5.3 Regular Security Updates

Frequent updates are crucial for maintaining the security and reliability of IoT devices and networks. It is important to regularly install the most recent security patches, firmware, and software updates on IoT devices to fix known vulnerabilities and bugs. Before implementing updates, thorough testing and verification should be conducted to prevent IoT security risks or any negative impact on device performance.

IoT devices should also be configured to automatically check for and download updates from the manufacturer's website or a trusted repository. Additionally, IoT devices should be configured to notify administrators and users of available updates and prompt them to install them at the earliest.

5.4 Network Segmentation

Network segmentation is an essential security measure that involves dividing a network into smaller subnetworks or segments to limit the spread of threats and reduce the impact of security breaches. Network segmentation allows organizations to group similar IoT devices and apply security policies based on the level of risk and criticality of each segment.

Network segmentation should be based on the principle of least privilege, where only

---

1 Two-factor authentication (2FA) is an identity and access management security method that requires two forms of identification to access resources and data.

2 A digital certificate is a file or electronic password that proves the authenticity of a device, server, or user through the use of cryptography and the public key infrastructure (PKI). Digital certificate authentication helps organizations ensure that only trusted devices and users can connect to their networks.

authorized devices and users are allowed to access specific segments. Segments should be isolated from each other to prevent lateral movement and propagation of threats. Additionally, IoT devices should be configured to use firewalls and intrusion detection systems (IDS)[1] to monitor and block suspicious traffic and activities.

6. Future of IoT Security

The exponential rise in the IoT has raised significant security concerns. With the increasing number of internet-connected devices, cyber criminals have a wider range of opportunities to exploit. This section will explore the prospective developments in IoT security and the emerging patterns expected to influence it.

6.1 Artificial Intelligence (AI) and Machine Learning (ML)

The future of IoT security is expected to witness a substantial contribution from advanced technologies like artificial intelligence (AI) and machine learning (ML). These innovative tools can swiftly identify and counter cyber threats to ensure prompt action to prevent major harm.

6.2 Blockchain Technology

The utilization of blockchain technology can improve the security and privacy of IoT devices. Through the implementation of blockchain, data can be stored in a decentralized and highly secure manner, thereby increasing the complexity of cyber criminals attempting to obtain unauthorized access to confidential information.

New Words

strategy	['strætədʒi]	n. 策略，战略
safe	[seɪf]	adj. 安全的
smoothly	['smuːðli]	adv. 平滑地；流畅地；平稳地
significance	[sɪɡ'nɪfɪkəns]	n. 重要性；意义
overstate	[ˌəʊvə'steɪt]	vt. 夸大（某事）；把……讲得过分；夸张
critical	['krɪtɪkl]	adj. 关键的；极重要的
breach	[briːtʃ]	n. 破坏；破裂；违背
		vt. 攻破；破坏，违反
expose	[ɪk'spəʊz]	v. 使暴露；使显露；揭露
detrimental	[ˌdetrɪ'mentl]	adj. 有害的；不利的
		n. 有害的人（或物）
consequence	['kɒnsɪkwəns]	n. 结果；后果
disruption	[dɪs'rʌpʃn]	n. 毁坏；中断
repercussion	[ˌriːpə'kʌʃn]	n. 反响，影响
threat	[θret]	n. 威胁
prevalent	['prevələnt]	adj. 普遍存在的，普遍发生的

1　An intrusion detection system (IDS) is a monitoring system that detects suspicious activities and generates alerts when they are detected.

attention	[ə'tenʃn]	n.	注意，注意力
botnet	['bɒtnet]	n.	僵尸网络
malware	['mælweə]	n.	恶意软件，流氓软件
infiltrate	['ɪnfɪltreɪt]	v.	（使）渗透，（使）渗入
vulnerable	['vʌlnərəbl]	adj.	易受攻击的
functionality	[ˌfʌŋkʃə'næləti]	n.	功能，功能性
tamper	['tæmpə]	v.	篡改
intercept	[ˌɪntə'sept]	vt.	拦截，拦住
confidential	[ˌkɒnfɪ'denʃl]	adj.	秘密的；机密的
hardship	['hɑːdʃɪp]	n.	艰难；困苦
adverse	['ædvɜːs]	adj.	不利的；有害的
default	[dɪ'fɔːlt]	n.	缺省，默认
usability	[ˌjuːzə'bɪlɪti]	n.	可用性；适用性
integrity	[ɪn'tegrətɪ]	n.	完整
target	['tɑːgɪt]	n.	目标，目的
instrumentation	[ˌɪnstrəmen'teɪʃn]	n.	使用仪器
visibility	[ˌvɪzə'bɪləti]	n.	可见性；能见度
remedial	[rɪ'miːdɪəl]	adj.	补救的，挽回的；纠正的
anomalous	[ə'nɒmələs]	adj.	不正常的；不协调的；不恰当的
novel	['nɒvl]	adj.	异常的
validate	['vælɪdeɪt]	vt.	使生效；批准，确认；证实
privilege	['prɪvəlɪdʒ]	n.	特权
attempt	[ə'tempt]	vt.	试图；尝试
traceability	[ˌtreɪsə'bɪləti]	n.	可追溯性，可追踪性
ciphertext	['saɪfətekst]	n.	密文
decrypt	[diː'krɪpt]	v.	解密，译（电文）
misuse	[ˌmɪs'juːz]	vt.	错用，滥用
regular	['regjələ]	adj.	有规律的，定期的
frequent	['friːkwənt]	adj.	频繁的
patch	[pætʃ]	n.	补丁，补片
vulnerability	[ˌvʌlnərə'bɪləti]	n.	弱点，脆弱性
bug	[bʌg]	n.	故障，程序错误，缺陷
verification	[ˌverɪfɪ'keɪʃn]	n.	证明；证实
download	[ˌdaʊn'ləʊd]	v.	下载
repository	[rɪ'pɒzətrɪ]	n.	仓库，储藏室
administrator	[əd'mɪnɪstreɪtə]	n.	管理者
measure	['meʒə]	n. & v.	测量，估量
subnetwork	[sʌb'netwɜːk]	n.	子网络，分网络

criticality	[krɪtɪ'kælɪti]	n. 危险程度
suspicious	[sə'spɪʃəs]	adj. 可疑的，不信任的
traffic	['træfɪk]	n. 流量，通信量
exponential	[ˌekspə'nenʃl]	n. 指数
		adj. 指数的，幂数的
significant	[sɪg'nɪfɪkənt]	adj. 重要的；显著的
opportunity	[ˌɒpə'tjuːnəti]	n. 机会
prospective	[prə'spektɪv]	adj. 预期的，未来的
influence	['ɪnfluəns]	n. 影响；势力
		vt. 影响；对……起作用
witness	['wɪtnəs]	n. 目击者，见证人
		vi. 做证人，见证
swiftly	[swɪftli]	adv. 迅速地，敏捷地
harm	[hɑːm]	n. & vt. 伤害，损害，危害
decentralize	[ˌdiː'sentrəlaɪz]	vt. 分散

Phrases

umbrella term	总称
user data	用户数据
cyber attack	网络攻击
identity theft	身份盗窃，身份盗用
essential infrastructure	基础设施
power grid	电网
unauthorized access	未经授权的访问，越权存取
cyber criminal	网络犯罪分子
engage in	参加，从事
break into	攻入
network security	网络安全
spread on	蔓延
device security	设备安全
internet-connected device	联网设备
mobile device management	移动设备管理
restricted network access	受限网络访问
device authentication	设备身份验证
authentication method	身份验证方法
biometric authentication	生物识别认证
two-factor authentication	双因素认证
digital certificate	数字证书

audit trail	审计跟踪
encryption algorithm	加密算法
encryption key	秘钥
security risk	安全风险
negative impact	负面影响
divide ... into ...	把……分为……
be isolated from ...	与……隔离

Abbreviations

DDoS (Distributed Denial of Service)	分布式拒绝服务
PC (Personal Computer)	个人计算机
ACL (Access Control List)	访问控制列表
AES (Advanced Encryption Standard)	高级加密标准
SHA (Secure Hash Algorithm)	安全哈希算法，安全散列算法
IDS (Intrusion Detection Systems)	入侵检测系统
ML (Machine Learning)	机器学习

Analysis of Difficult Sentences

[1] IoT security is an umbrella term for the tools and strategies that protect devices connected to the cloud and the network they use to connect to each other.

本句中，that protect devices connected to the cloud and the network they use to connect to each other 是一个定语从句，修饰和限定 the tools and strategies。在该从句中，connected to the cloud 是一个过去分词短语，作后置定语，修饰和限定 devices，它可以扩展为一个定语从句：which are connected to the cloud。they use to connect to each other 是一个定语从句，修饰和限定 the network。

[2] Any breach in security may expose this data, leading to detrimental consequences like identity theft and financial harm.

本句中，leading to detrimental consequences like identity theft and financial harm 是一个现在分词短语，作结果状语。lead to 的意思是"导致，引起，造成"。

[3] IoT devices are often vulnerable to malware attacks, which can compromise their functionality and steal data.

本句中，which can compromise their functionality and steal data 是一个非限定性定语从句，对 malware attacks 进行补充说明。be vulnerable to 的意思是"易受"。例如：

Applications can be vulnerable to two kinds of security threats: dynamic and static.
有两类攻击可能会威胁应用程序的安全性：静态威胁和动态威胁。

All of these are points which could be vulnerable to attacks by hackers and other intruders.
这些都是易受黑客及其他入侵者攻击的漏洞。

[4] Weak passwords are a common IoT security threat, as many IoT devices use default passwords that are easy to guess or crack.

本句中，as many IoT devices use default passwords that are easy to guess or crack 是一个原因状语从句。在该从句中，that are easy to guess or crack 是一个定语从句，修饰和限定 default passwords。

[5] Additionally, IoT devices should be configured to use secure communication protocols such as transport layer security (TLS) or secure sockets layer (SSL) to protect data that is transmitted over the internet.

本句中，such as transport layer security (TLS) or secure sockets layer (SSL)是对 secure communication protocols 的举例说明。to protect data that is transmitted over the internet 是动词不定式短语，作目的状语。在该短语中，that is transmitted over the internet 是一个定语从句，修饰和限定 data。

[6] Network segmentation is an essential security measure that involves dividing a network into smaller subnetworks or segments to limit the spread of threats and reduce the impact of security breaches.

本句中，that involves dividing a network into smaller subnetworks or segments to limit the spread of threats and reduce the impact of security breaches 是一个定语从句，修饰和限定 an essential security measure。在该从句中，to limit the spread of threats and reduce the impact of security breaches 是动词不定式短语，作目的状语。

Exercises

【EX.1】Fill in the following blanks according to the text.

1. The main goals of IoT security are to _____, _____ and _____.
2. IoT devices are crucial to essential infrastructure, such as _____, _____, and _____.
3. Botnets are networks of compromised devices that cyber criminals can controll to launch _____, _____, or engage in other malicious activities.
4. IoT devices frequently retain confidential information, including _____, _____, and _____.
5. There are three distinct types of IoT security. They are _____, _____, and _____.
6. Network security is any activity designed to protect the _____ and _____ of your network and data. It includes both _____ and _____ security.
7. Device authentication validates the identity of _____, which permits network access solely to _____. Access control, in turn, governs _____ granted to individual devices or users.
8. Encryption entails using _____ and _____ to convert plain text data into _____.
9. Network segmentation should be based on the principle of _____, where only authorized devices and users are allowed to _____.
10. Through the implementation of blockchain, data can be stored in a _____ and _____ manner, thereby increasing the complexity of cyber criminals attempting to obtain unauthorized access to _____.

【EX.2】 Translate the following terms or phrases from English into Chinese and vice versa.

1. blockchain   1. _____
2. breach   2. _____
3. bug   3. _____
4. ciphertext   4. _____
5. default   5. _____
6. functionality   6. _____
7. integrity   7. _____
8. opportunity   8. _____
9. authentication method   9. _____
10. device authentication   10. _____
11. 数字证书   11. _____
12. 秘钥   12. _____
13. 网络安全   13. _____
14. *n.* 流量，通信量   14. _____
15. *v.* 篡改   15. _____

【EX.3】 Translate the following sentences into Chinese.
1. We take threats of this kind very seriously.
2. Protecting yourself from identity theft is a matter of treating all your personal and financial documents as top secret information.
3. The server is designed to store huge amounts of data.
4. The company suffered a major cyber attack in which thousands of files were taken by hackers.
5. Malware detection is an important area in the computer security.
6. You can download the file and edit it on your word processor.
7. The agent cannot decrypt the message with the public key.
8. It is significantly more compact than any comparable laptop, with no loss in functionality.
9. The default is usually the setting that most users would probably choose.
10. The bug was caused by an error in the script.

【EX.4】 Complete the following passage with appropriate words in the box.

| constraints | update | inventor | basic | symbol |
| interactive | mainstream | descriptions | scan | equal |

  Barcodes are a pretty big deal – and not just at the grocery store. 2D barcodes are the ___1___ of recognition technologies used for mobile marketing. You'll be hearing more about newcomers such as NFC, but for now, the majority of folks are sticking with 2D.

In mobile tagging, the barcode is a printed ___2___ that connects a physical object (a magazine ad) to a digital experience on a smartphone (a cool video). Why should you care? Because a 2D barcode like a Microsoft Tag barcode adds a whole new dimension to your marketing campaigns, making them more engaging and ___3___.

You can put a 2D barcode on just about anything – printed materials, packaging, posters, signs, websites, clothing. When people can ___4___ the barcode with their smartphones, they instantly see the online content you've created – from a product video to a sweepstakes to a custom mobile site.

But not all barcodes are created ___5___. The type of barcode you use is important, because features and ease of use vary. There are three types of barcodes in common use: Microsoft Tag barcodes, QR codes, and traditional linear barcodes.

1. Tag Barcodes

Tag barcodes are the newest edition of 2D barcodes. They offer more flexibility than older formats both in the barcode design and the content behind it. Because tag barcodes are linked to data stored on a server, you can deliver a more robust online experience – including entire mobile sites – and ___6___ the content any time without having to change the Tag. So, if you link a tag on your business card to your resume, it will still be valid after you get that big promotion. Tags can be black-and-white or full-color, including custom images (e.g., a company logo).

2. QR Code

The quick response (QR) code was the earliest 2D barcode. It was designed to be a bump up from its predecessor, the 1D barcode, because it can contain more information. While not technically open source, the ___7___ of the QR code and owner of the QR code trademark, DENSO, has allowed the patents for the code to be freely available to the public. QR Codes have a variety of disparate formats and reader Apps, and can be black-and-white or basic colors. Because of these ___8___, QR codes are best suited for simple designs that don't require integration with your branding.

3. Traditional Barcodes

It's not likely, but it's possible the 1D barcode on your loaf of bread carries a little something extra. Some marketers provide ___9___ product information using the 1D barcodes you've known for years. Some services use mobile apps to scan these barcodes and display data such as prices, ___10___, and user reviews.

【EX.5】 Translate the following passage into Chinese.

What Is 2D Barcode?

A 2D (two-dimensional) barcode is a graphical image that stores information both horizontally and vertically. As a result, 2D codes can store up to 7089 characters, significantly greater storage than is possible with the 20-character capacity of a unidimensional barcode.

2D barcodes are also known as quick response codes because they enable fast data access. 2D barcodes are often used in conjunction with smart phones. The user simply photographs a 2D barcode with the camera on a phone equipped with a barcode reader. The reader interprets the encoded URL, which directs the browser to the relevant information on a website. This capability has made 2D barcodes useful for mobile marketing. Some 2D barcode systems also deliver information in a message for users without Web access.

Here are some examples how 2D barcodes are being used:
- Nike used 2D barcodes on posters along the route of an extreme sports competition. Mobile users captured barcodes to access sponsored pictures, video and data.
- Some newspapers include 2D barcodes on stories that link mobile users to developing coverage.
- 2D barcodes on products in stores link to product reviews.
- Some people post 2D barcodes that link to their blogs or Facebook pages.

Text B

Blockchain for IoT

Blockchain is a distributed ledger technology[1] that enables secure and transparent record-keeping of transactions and data transfers.Blockchain for IoT refers to the use of blockchain technology in IoT devices and networks. By using blockchain in IoT, devices can securely communicate and transact with each other without the need for a centralized intermediary.This can improve the security and privacy of IoT networks.

Blockchain can provide several benefits for IoT, including enhanced security, data integrity[2], and decentralized control. By using blockchain, IoT devices can create an immutable record of all transactions and data transfers, making it difficult for hackers to tamper with or manipulate data. Blockchain can also provide a decentralized control structure, allowing IoT networks to operate without a single point of failure.

1. How Blockchain Can Enhance IoT Security

Decentralization: blockchain is a decentralized technology, which means that there is no central authority or control over the data stored on the blockchain. This makes it difficult for hackers to attack a single point of failure in the network, making the IoT ecosystem more secure.

---

1 Distributed ledger technologies, like blockchain, are peer-to-peer networks that enable multiple members to maintain their own identical copy of a shared ledger. Rather than requiring a central authority to update and communicate records to all participants, DLTs allow their members to securely verify, execute, and record their own transactions without relying on a middleman.

2 Data integrity is the overall accuracy, completeness, and consistency of data. Data integrity also refers to the safety of data in regard to regulatory compliance — such as GDPR compliance — and security. It is maintained by a collection of processes, rules, and standards implemented during the design phase.

Immutable and tamper-proof: once data is recorded on the blockchain, it cannot be altered or deleted. This ensures that the integrity of the data is maintained, and any attempts to tamper with the data are immediately detected.

Smart contracts: smart contracts are self-executing programs that automatically enforce the terms of an agreement between two parties. By using smart contracts on a blockchain, IoT devices can autonomously execute transactions and enforce rules without the need for a central authority, reducing the risk of fraud or errors.

Access control[1]: blockchain can be used to implement access control mechanisms that limit who can access and modify data. This can help prevent unauthorized access to IoT devices and data.

Transparency: the transparency of blockchain technology allows for greater visibility into the IoT ecosystem. By providing a tamper-proof audit trail of all transactions, blockchain can enable more effective monitoring and analysis of IoT data, making it easier to detect anomalies and potential security threats.

2. IoT Security Challenges

The security of IoT devices is facing numerous challenges due to the diverse nature of the ecosystem. One of the major challenges is the lack of standardization as IoT devices often come from different manufacturers and may use different communication protocols and security mechanisms. This makes it difficult to ensure consistent security across the ecosystem.

Another challenge is the weak passwords and authentication mechanisms of many IoT devices. Users may not take the time to change default or weak passwords, making it easy for hackers to gain access to the devices and data. Additionally, IoT devices often run on software that may not receive regular security updates, leaving them vulnerable to known and unknown security threats. The collection and transmission of large amounts of data by IoT devices also raises concerns about data privacy and confidentiality.

Many IoT devices lack adequate security measures to protect against data breaches or leaks, putting sensitive and personal information at risk. Moreover, IoT devices can be used as part of botnets to launch DDoS attacks, which can bring down websites and other online services. Finally, many IoT devices are deployed in public or unsecured locations, making them vulnerable to physical attacks or theft. Addressing these challenges requires a concerted effort from all stakeholders, including manufacturers, developers, users, and policymakers, to ensure that IoT devices are secure by design and that proper security measures are in place to protect them throughout their lifecycle.

---

1 Access control is a data security process that enables organizations to manage who is authorized to access corporate data and resources. Secure access control uses policies that verify users are who they claim to be and ensures appropriate control access levels are granted to users.

3. Benefits of Using Blockchain Technology for IoT

Enhanced security: by ensuring that data exchanged between devices is encrypted and tamper-proof, blockchain technology can give IoT devices enhanced security. The decentralized nature of blockchain technology also makes it more difficult for hackers to attack the network because they would have to simultaneously compromise several network nodes.

Improved transparency: the network's transactions and data transfers are all recorded transparently and with the ability to be audited using blockchain technology. As a result, the IoT ecosystem can become more transparent and accountable because everyone can see the same data in real time.

Increased efficiency: by offering a decentralized and automated method of managing data transfers and transactions between devices, blockchain technology can assist to increase the efficiency of IoT networks. This may lessen the need for middlemen and increase the efficiency and precision of data transmissions.

Cost-effective: blockchain technology can assist to bring down the overall expenses of administering IoT networks by doing away with the need for middlemen and centralized data storage solutions. Small and medium-sized enterprises, which might not have the capacity to invest in costly infrastructure, can particularly benefit from this.

Scalability: blockchain technology can handle enormous amounts of data transfers and transactions and is very scalable. This makes it the perfect answer for the IoT ecosystem, which is predicted to grow rapidly over the coming years and produce billions of connected devices and enormous amounts of data.

4. Things to Know about Blockchain Applications in IoT

The application of blockchain in IoT offers various advantages that can significantly improve the efficiency and security of IoT networks. Firstly, it can enhance security by providing a decentralized, tamper-proof ledger of all transactions and data transfers, thereby preventing unauthorized access and data breaches. Blockchain technology can also ensure data integrity by providing a transparent and immutable record of all transactions, preventing data tampering and other types of data manipulation. In addition, blockchain-based smart contracts can automate many IoT processes, leading to faster and more efficient data transfers and other operations.

Blockchain can decentralize IoT networks, reducing the reliance on centralized servers and other infrastructure, which can improve network resilience and reduce the risk of single points of failure. Tokenization[1], which is a process of creating digital assets on the blockchain, can incentivize IoT device owners and users, enabling new business models and revenue streams. Additionally, blockchain-based IoT solutions have a wide range of potential applications in industries such as healthcare, supply chain management, logistics, energy, and more.

---

1 Tokenization is the process of exchanging sensitive data for nonsensitive data called "tokens" that can be used in a database or internal system without bringing it into scope.

5. Challenges and Limitations of Blockchain for IoT

Blockchain has the potential to enhance the security and efficiency of IoT networks, but there are several challenges and limitations that need to be addressed. Scalability is a major challenge as IoT networks generate vast amounts of data and blockchain-based solutions may struggle to keep up. Energy consumption is also a concern and requires the development of energy-efficient consensus algorithms and hardware solutions. Interoperability is crucial for blockchain-based IoT systems to be effective and requires standardization of communication protocols and data formats. Implementing blockchain-based solutions can be expensive and subject to regulatory compliance. User adoption is critical as many users may not be familiar with blockchain technology. Finally, while blockchain can enhance security, it is not a solution for all security threats and there are still vulnerabilities and risks associated with blockchain-based IoT systems.

6. Future Outlook and Potential for Blockchain for IoT

The future outlook for blockchain for IoT is promising, as the combination of these two technologies has the potential to revolutionize the way devices and networks interact and operate. The decentralized and tamper-proof nature of blockchain can enhance the security and efficiency of IoT networks, while also enabling new business models and revenue streams.

One potential application of blockchain for IoT is in the area of supply chain management, where blockchain-based systems can provide greater transparency and traceability of goods and materials. This can help prevent counterfeiting and fraud, and ensure the integrity of supply chain data.

Another potential application of blockchain in IoT is in the area of energy management, where blockchain-based systems can enable more efficient and decentralized energy production and distribution. This can help reduce energy waste and costs, while also improving the reliability and resilience of energy networks.

Blockchain-based IoT solutions also have potential applications in healthcare, where they can enable secure and efficient sharing of patient data and medical records between healthcare providers and patients. This can help improve patient outcomes and reduce healthcare costs.

New Words

blockchain	[blɒkt'ʃeɪn]	n.	区块链
record-keeping	['rekɔːd'kiːpɪŋ]	n.	记录保存
centralize	['sentrəlaɪz]	vt.	使集中，中心化
immutable	[ɪ'mjuːtəbl]	adj.	不可改变的
manipulate	[mə'nɪpjuleɪt]	vt.	操作，处理
decentralization	[ˌdiːˌsentrəlaɪ'zeɪʃn]	n.	分散，去中心化
authority	[ɔː'θɒrəti]	n.	权威，权力
tamper-proof	['tæmpəpruːf]	n.	防干扰

alter	[ˈɔːltə]	v. 改变，更改
delete	[dɪˈliːt]	v. 删除
self-executing	[ˈself ˈeksɪkjuːtɪŋ]	adj. 自动生效的
agreement	[əˈɡriːmənt]	n. 协定，协议
transparency	[trænsˈpærənsi]	n. 透明，透明度，透明性
confidentiality	[ˌkɒnfɪˌdenʃɪˈæləti]	n. 机密性
adequate	[ˈædɪkwət]	adj. 足够的；适当的
policymaker	[ˈpɒləsɪmeɪkə]	n. 政策制定者，决策人
lifecycle	[ˈlaɪfˌsaɪkl]	n. 生命周期
accountable	[əˈkaʊntəbl]	adj. 负有责任的
middleman	[ˈmɪdlmæn]	n. 中间人；经纪人
scalable	[ˈskeɪləbl]	adj. 可升级的
resilience	[rɪˈzɪlɪəns]	n. 弹性，弹力
tokenization	[təʊˈkɪnaɪzeɪʃn]	n. 令牌化，标记化
incentivize	[ɪnˈsentɪvaɪz]	vt. 激励
logistic	[ləˈdʒɪstɪkl]	n. 物流
struggle	[ˈstrʌɡl]	vi. 努力；争取
interoperability	[ˈɪntərɒpərəˈbɪləti]	n. 互用性
regulatory	[ˈreɡjələtəri]	adj. 监管的
promising	[ˈprɒmɪsɪŋ]	adj. 有前途的，有希望的
interact	[ˌɪntərˈækt]	v. 相互作用，互相影响
counterfeit	[ˈkaʊntəfɪt]	n. 仿制品
		v. 仿制，造假

Phrases

distributed ledger technology	分布式账本技术
data integrity	数据完整性
decentralized control	分散控制，非集中控制
tamper with	损害；篡改
decentralized control structure	分散控制结构，去中心化控制结构
smart contract	智能合约
automatically enforce	自动执行
online service	联机服务，在线服务
unsecured location	不安全的位置，不安全的地方
small and medium-sized enterprise	中小企业
immutable record	不可变记录
data manipulation	数据操作
revenue stream	收入来源

supply chain management　　　　　　　供应链管理
blockchain-based solution　　　　　　　基于区块链的解决方案
energy management　　　　　　　　　　能源管理
energy waste　　　　　　　　　　　　　能量损耗，能量浪费

Exercises

【EX.6】 Answer the following questions according to the text.

1. What is blockchain?
2. What does blockchain is a decentralized technology mean?
3. What are smart contracts?
4. What is one of the major challenges the security of IoT devices is facing? Why?
5. Why does the decentralized nature of blockchain technology also make it more difficult for hackers to attack the network?
6. How can the application of blockchain in IoT enhance security?
7. What are the industries that blockchain-based IoT solutions have a wide range of potential applications?
8. Why is scalability a major challenge of blockchain for IoT?
9. What can the decentralized and tamper-proof nature of blockchain do?
10. What is one potential application of blockchain for IoT?

【EX.7】 Translate the following terms or phrases from English into Chinese and vice versa.

1. agreement　　　　　　　　　　1. _____
2. centralize　　　　　　　　　　2. _____
3. delete　　　　　　　　　　　　3. _____
4. interoperability　　　　　　　　4. _____
5. manipulate　　　　　　　　　　5. _____
6. regulatory　　　　　　　　　　6. _____
7. resilience　　　　　　　　　　　7. _____
8. transparency　　　　　　　　　8. _____
9. data manipulation　　　　　　　9. _____
10. energy waste　　　　　　　　　10. _____
11. 供应链管理　　　　　　　　　　11. _____
12. 能源管理　　　　　　　　　　　12. _____
13. 分散控制，非集中控制　　　　　13. _____
14. 联机服务，在线服务　　　　　　14. _____
15. 数据完整性　　　　　　　　　　15. _____

【EX.8】 **Translate the following sentences into Chinese.**
1. The data integrity is an important content of the database protection.
2. This is the entire history of commerce on that blockchain.
3. The computer is programmed to warn users before information is deleted.
4. User input may be used to adjust filtering and tokenization of the messages.
5. Early in the development lifecycle, one uses simulation and analysis tools.
6. They are trying to improve the service management for firm's logistic?
7. Like other counterfeits, they look like real products.
8. Computers are very efficient at manipulating information.
9. All information received from you will be treated with strict confidentiality.
10. Millions of people want new, simplified ways of interacting with a computer.

Reading Material

Cloud IoT

From power grids and telecommunications networks, to rail, sea, and air transport systems, infrastructure today uses millions of sensors. Gathering data from all of these sensors and using it to ensure the smooth functioning of this critical infrastructure is made possible by combining cloud computing and the IoT.

Cloud IoT uses cloud computing services to collect and process data from IoT devices, and to manage the devices remotely. The scalability of cloud IoT platforms enables the processing of large amounts of data, as well as artificial intelligence (AI) and analytics capabilities.

1. What Is Cloud IoT?

Cloud IoT is a technology architecture that connects IoT devices to servers housed in cloud data centers. This enables real-time data analytics, allowing better, information-driven[1] decision making, optimization and risk mitigation[2]. Cloud IoT also simplifies management of connected devices at-scale.

Cloud IoT is different from traditional, or non-cloud-based IoT in a few key ways:
- Data storage: the cloud collects IoT data generated by thousands or millions of IoT sensors, with the data being stored and processed in a central location. While in other types of IoT architectures, data may be stored and processed on-premises[3].
- Scalability: cloud IoT is highly scalable, as cloud infrastructure (compute, storage, and networking resources) can easily handle thousands of devices and process their data across large systems.
- Flexibility: cloud IoT provides a high level of flexibility, as it allows devices to be added

---

1 information-driven [ˌɪnfəˈmeɪʃn ˈdrɪvn] *adj.* 信息驱动的
2 mitigation [ˌmɪtɪˈgeɪʃn] *n.* 缓解，减轻
3 on-premises [ɒn- ˈpremɪsɪz] *n.* 现场

or removed as needed, without having to reconfigure[1] the entire system.
- Maintenance: in cloud IoT, the maintenance of servers and networking equipment is handled by the cloud service provider (CSP)[2]. While in other types of IoT architectures, maintenance may be the responsibility[3] of the end user.
- Cost: cloud IoT can be more cost-effective over the long-term, as users only pay for the resources they actually consume, and users do not have to invest upfront in their own expensive compute, storage, and networking infrastructure.

2. How Does Cloud IoT Work?

Cloud IoT connects IoT devices—which collect and transmit data—to cloud-based servers via communication protocols such as MQTT and HTTP and over wired and wireless networks. These IoT devices can be managed and controlled remotely and integrated with other cloud services.

IoT data is sourced from anywhere and everywhere, including sensors, actuators, operating systems, mobile devices, standalone applications, and analytic systems. By involving the cloud, vast amounts of IoT data can be stored and processed in a central location.

A cloud IoT system typically includes the following elements:

- IoT devices: physical devices, such as sensors and actuators, that generate and transmit data to the cloud.
- Connectivity: communication protocols and standards used to connect the IoT devices to the cloud. Examples of protocols include MQTT and HTTP, while examples of standards are WiFi, 4G, 5G, Zigbee, and LoRa (long range).
- Cloud platforms: cloud service providers that offer infrastructure and services to connect to the IoT devices. Examples include AWS IoT and Azure IoT.
- Data storage: cloud-based storage for data generated by the IoT devices, which can be housed in repositories such as a database, data warehouse[4], or data lake[5].
- Application layer or API: cloud IoT platforms typically provide a native[6] application — for analytics, machine learning (ML), and visualization — or application programming interface (API) — for data processing. Usually, applications offer the ability to manage and monitor the IoT devices for provisioning, software updates, and troubleshooting[7].
- Security: measures put in place to secure the data and IoT devices, such as encryption, authentication, and access control.

1　reconfigure [ˌriːkənˈfɪɡə] v. 重新装配
2　cloud service provider (CSP)：云服务提供商
3　responsibility [rɪˌspɒnsəˈbɪləti] n. 责任，职责
4　data warehouse：数据仓库
5　data lake：数据湖
6　native [ˈneɪtɪv] adj. 本地的
7　troubleshooting [ˈtrʌblʃuːtɪŋ] n. 发现并修理故障

3. What Are the Cloud Services for IoT?

Cloud platforms deliver a collection of capabilities that allow IoT devices to interact with cloud services, other applications and even other IoT devices. These cloud platforms let users centrally onboard[1], manage, monitor, and control IoT devices.

In addition, the cloud supports services such as scalable storage, device connectivity, analytics and reporting, and identity and access management (IAM)[2] in IoT.

3.1 Scalable Storage

Cloud IoT platforms provide scalable object storage services, such as Amazon simple storage service (Amazon S3), that allow organizations to easily increase or decrease their data storage requirements. This type of flexibility is beneficial for IoT applications, as they often generate large volumes of unstructured data and must be able to store this information without sacrificing[3] device performance.

3.2 Device Connectivity

Cloud-based IoT platforms offer straightforward, reliable, and secure connectivity at scale between physical IoT devices and cloud services. In turn, an organization can connect thousands or millions of IoT devices to the cloud, without the need to provision or manage the requisite servers and networking equipment.

3.3 Analytics and Reporting

Cloud-based IoT platforms are equipped with powerful analytics capabilities—in combination with computing resources — that enable organizations to gain real-time insights into the large datasets that IoT devices produce. Through sophisticated algorithms, such as predictive modeling, statistical analysis, and machine learning (ML), IoT device data can be used to improve efficiency and make better, information-driven decisions.

Additionally, IoT device data can be combined with[4] other relevant data stored in the cloud cloud to extract meaningful insights for organizations. Furthermore, built-in data reporting features offered by cloud services allow organizations to create useful reports based on collected IoT data.

3.4 Identity and Access Management (IAM)

Security for the data generated by IoT devices can be protected in the cloud using identity and access management (IAM), which is an authentication and authorization service. IAM enables organizations to grant or deny access[5] to services and resources in the cloud for large numbers of users with different access needs.

With so much IoT data being sent to the cloud, the granularity[6] of IAM controls allows

---

1 onboard ['ɒn'bɔːd] v. 搭载
2 identity and access management (IAM)：身份和访问管理
3 sacrifice ['sækrɪfaɪs] n. & v. 牺牲
4 be combined with：与……结合
5 deny access：拒绝访问
6 granularity [ˌɡrænjʊ'lærɪti] n. 粒度

organizations to comply with security and regulations that are relevant for storing and accessing sensitive information.

参考译文

Text A 物联网安全

1. 什么是物联网安全？

物联网安全是保护连接到云的设备及其用于相互连接的网络的工具和策略的总称。其主要目标是保护用户数据安全、阻止网络攻击并保持设备平稳运行。

2. 物联网安全的重要性

由于各种因素，物联网安全的重要性怎么强调都不为过。物联网设备通常存储必须受到保护的关键信息，如财务数据和个人数据。任何安全漏洞都可能会暴露这些数据，从而导致身份盗窃和财务损失等不利后果。

物联网设备对于电网、交通系统和医疗卫生等重要的基础设施来说至关重要。未经授权访问这些系统可能导致严重后果，如电力故障、交通中断及潜在的生命损失。

3. 物联网安全威胁

物联网安全面临着各种不断变化的风险。以下是一些需要关注的常见物联网安全威胁。

- 僵尸网络：僵尸网络是由受感染设备组成的网络，网络犯罪分子可以控制这些设备，并用其发起分布式拒绝服务（DDoS）攻击、窃取数据或从事其他恶意活动。由于物联网设备数量大、安全性弱，因此经常被用于僵尸网络。
- 恶意软件：恶意软件是旨在渗透和损坏计算机及其他设备的软件。物联网设备通常易受恶意软件的攻击，这可能会损害其功能并且造成数据被窃取。
- 物理篡改：可以对物联网设备进行物理篡改，以获得对设备或网络的未经授权的访问。这可能涉及侵入设备本身或拦截设备与网络之间的信号。
- 数据泄露：物联网设备经常保留机密信息，包括个人数据、财务记录和其他敏感详细信息。如果存在安全漏洞，那么这些数据的暴露可能会导致严重的后果，如身份盗窃、财务困难和其他不利后果。
- 弱密码：弱密码是常见的物联网安全威胁，因为许多物联网设备都使用易于猜测或破解的默认密码，这使得攻击者可以轻松访问物联网设备和网络。

4. 物联网安全的类型

物联网安全分为以下三种类型。

- 网络安全：网络安全是旨在保护网络和数据的可用性和完整性的任何活动。它包括硬件安全和软件安全，目标是应对各种威胁。它可以阻止他人进入你的网络或在你的网络上传播信息。有效的网络安全可以管理对网络的访问。
- 设备安全：设备安全指保护移动电话、笔记本电脑、个人计算机、平板电脑、物联网设备等联网设备免受网络威胁和未经授权的访问。设备安全通常需要强大的身份验证、移动设备管理工具和受限的网络访问。
- 固件安全：固件安全首先要深入检查物联网设备的固件，以检测其代码中的潜在漏洞。固件安全使用仪器来监控设备上运行的二进制文件的状态。这种更深入的可见性使固件安全解决方案能够根据受监控的二进制文件的异常行为来识别欲攻击行为和修复损坏的固件，并识别和防止新的物联网漏洞。

5. 物联网安全的最佳实践

5.1 设备认证与访问控制

确保物联网的安全涉及两个基本要素，即设备身份验证和访问控制。设备身份验证可验证物联网设备的身份，仅允许授权的设备访问网络。访问控制对于授予单个设备或用户的权限范围加以管理。为了增强安全性，物联网设备应配置采用强大的身份验证方法，如生物识别身份验证、双因素身份验证或数字证书，以验证设备和用户的身份。

还应实施访问控制，根据每个设备或用户的角色和信任级别来限制其访问权限。这可以通过使用访问控制列表(ACL)来实现，该列表指定设备或用户访问特定资源或执行某些操作的权限。此外，物联网设备应配置为能够记录所有访问尝试和活动，以实现审计跟踪和可追溯性。

5.2 加密

加密是物联网安全的重要组成部分，可确保物联网设备和网络之间传输的数据的机密性和完整性。加密需要使用加密算法和密钥将纯文本数据转换为密文。只有拥有正确密钥的授权用户才能解密密文并访问明文数据。

物联网设备应使用高级加密标准（AES）或安全哈希算法（SHA）等强大的加密算法来保护传输中的数据和静态数据的安全。此外，物联网设备应配置为使用安全通信协议，如传输层安全性（TLS）或安全套接字层（SSL），来保护通过互联网传输的数据。应安全地存储和管理加密密钥，以防止未经授权的访问和滥用。

5.3 定期安全更新

频繁更新对于维护物联网设备和网络的安全性和可靠性至关重要。定期在物联网设备上安装最新的安全补丁、固件和软件更新以修复已知的漏洞和错误非常重要。在实施更新之前，应进行彻底的测试和验证，以防止物联网安全风险或对设备性能产生任何负面影响。

物联网设备还应该配置为自动检查并从制造商的网站或受信任的存储库下载更新。此外，物联网设备应配置为通知管理员和用户可用的更新，并提示他们尽早安装。

5.4 网络分段

网络分段是一项重要的安全措施，包括将网络划分为更小的子网或网段，以限制威胁的传播并减少安全漏洞的影响。网络分段允许组织对类似的物联网设备进行分组，并根据每个分段的风险和关键程度采用不同的安全策略。

网络分段应基于最小权限原则，仅允许授权设备和用户访问特定网段。各个网段应相互隔离，以防止威胁的横向移动和传播。此外，物联网设备应配置为使用防火墙和入侵检测系统(IDS)来监控和阻止可疑流量和活动。

6. 物联网安全的未来

物联网的指数级增长已经引发了重大的安全问题。随着联网设备数量的不断增加,网络犯罪分子有了更广泛的利用机会。下面将探讨物联网安全的未来发展及预计会影响它的新兴模式。

6.1 人工智能(AI)和机器学习(ML)

物联网安全的未来预计将见证人工智能(AI)和机器学习(ML)等先进技术的重大贡献。这些创新工具可以快速识别和应对网络威胁,以确保及时采取行动,防止重大损害。

6.2 区块链技术

利用区块链技术可以提高物联网设备的安全性和隐私性。通过实施区块链技术,数据能够以分散且高度安全的方式存储,从而增加了网络犯罪分子试图未经授权访问机密信息的复杂性。

Unit 10

Text A

扫码听音频

IoT Data Analytics

1. What Is IoT Data Analytics?

As its name suggests, IoT data analytics is the act of analyzing data generated and collected from IoT devices by utilizing a specific set of data analytics tools and techniques. The true idea behind IoT data analytics is to turn vast quantities of unstructured data[1] from various devices and sensors within the IoT ecosystem, which is heterogeneous, into valuable and actionable insights for driving sound business decision-making and further data analysis. Furthermore, IoT analytics enables identifying the patterns in data sets, including both current states and historical data, which can be utilized to make predictions and adjustments about future events.

2. Different Types of IoT Data Analytics

As IoT analytics are performed to gather insights that serve different purposes, it can be broken down into four primary types:

2.1 Descriptive Analytics

Descriptive IoT analytics mainly focus on what happened in the past. The historical data collected from devices are processed and analyzed to generate a report that describes what took place, when it occurred, and how often it did. This type of IoT analysis is useful for providing answers to specific questions about the behavior of things or people and can also be used to detect any anomalies.

---

1 Unstructured data (or unstructured information) is information that either does not have a pre-defined data model or is not organized in a pre-defined manner. Unstructured information is typically text-heavy, but may contain data such as dates, numbers, and facts as well. This results in irregularities and ambiguities that make it difficult to understand using traditional programs as compared to data stored in fielded form in databases or annotated (semantically tagged) in documents.

2.2 Diagnostic Analytics

Different from descriptive IoT analytics, diagnostic analytics go one step further to answer the question of why something happened by drilling down into the data to identify the root cause of a specific issue. Diagnostic analytics make use of techniques like data mining[1] and statistical analysis to uncover hidden patterns and relationships in data that can offer actionable insights into the causes of specific problems.

2.3 Predictive Analytics

As its name suggests, predictive IoT analytics is used to predict future events by analyzing historical data and trends. This type of analytics makes use of various statistical and machine learning algorithms to build models that can be used for making predictions about future events. This type of analytics plays a significant role in supporting business decisions related to inventory management, demand forecasting, etc.

2.4 Prescriptive Analytics

Prescriptive IoT analytics is the most advanced type of IoT analytics that not only predicts what will happen in the future but also provides recommendations on what should be done to achieve the desired business outcomes. This type of analytics makes use of optimization algorithms to identify the best course of action that should be taken to achieve a specific goal.

3. The Relation Between IoT and Big Data[2] Analytics

Speaking of massive amounts of data, you are being reminded of big data analytics, aren't you? Do they have any sort of connections? Actually, people often find IoT and big data analytics are confused by each other. The only distinction between them is the data source; while big data analytics deals with data sets from a broad range of streams and sources, IoT analytics only collect and analyze data generated by connected IoT devices and sensors. So, we can say that IoT data analytics is a subset of big data analytics that helps make sense of data originating from connected devices in the ecosystem of the IoT. And as a result, IoT analytics can be used to solve various issues and problems that cannot be addressed by big data analytics alone, such as real-time streaming data analysis, near-time processing, edge computing[3], predictive maintenance, etc. Therefore, the combination of IoT and big data analytics can be used to gain a competitive edge and drive business value.

4. Benefits of IoT Analytics

The benefits of IoT analytics are numerous, and they can be classified into two main

---

1 Data mining is the process of searching and analyzing a large batch of raw data in order to identify patterns and extract useful information.

2 Big data is a combination of structured, semistructured and unstructured data collected by organizations that can be mined for information and used in machine learning projects, predictive modeling and other advanced analytics applications.

3 Edge computing is an emerging computing paradigm which refers to a range of networks and devices at or near the user. Edge is about processing data closer to where it's being generated, enabling processing at greater speeds and volumes, leading to greater action-led results in real time.

categories: business benefits and technical benefits. Let's have a look at each of them in detail.

4.1 Business Benefits of IoT Analytics

- Optimizing operational efficiency: by analyzing data generated by IoT devices, businesses can identify issues and problems that lead to inefficiencies and then take actions to address them. For instance, a food & beverage company can use IoT data analytics to track the temperature of its refrigerators in real-time and prevent food spoilage due to power outages or malfunctioning equipment.
- Reducing costs: IoT data analytics can help businesses save money in many ways, such as reducing energy consumption, minimizing downtime, and improving asset utilization. For example, a manufacturing company can use IoT data analytics to monitor the performance of its production line and make adjustments to avoid wastage of materials.
- Enhancing customer experience: IoT data analytics can be used to collect and analyze customer data in order to understand their needs and preferences. This, in turn, can help businesses design better products and services that meet the needs of their customers. For instance, a retailer can use IoT data analytics to track the movements of customers in its store and then offer them personalized recommendations based on their interests.
- Improving safety: by analyzing data from various sensors, businesses can identify potential safety hazards and take preventive measures to avoid them. For instance, a construction company can use IoT data analytics to monitor the condition of its equipment and machinery in order to avoid accidents.

4.2 Technical Benefits of IoT Analytics

- Real-time data analysis: one of the main advantages of IoT data analytics is its ability to analyze real-time data points. It is possible due to the use of streaming analytics, which is a type of analytics that can process data as it is being generated.
- Improved scalability: with IoT data analytics, businesses can scale up their operations quickly and easily without incurring any additional costs. This is because IoT data analytics can be deployed on the cloud, which allows businesses to pay only for the resources they use.
- Increased accuracy: another advantage of IoT data analytics is that it can help businesses achieve a high degree of accuracy in their data analysis. This is due to the fact that IoT data analytics can be used to collect data from a large number of sources and then analyze it using advanced analytical techniques.
- Enhanced security: IoT data analytics can also help businesses improve the security of their data. This is because IoT data analytics can be used to identify and track potential threats and then take measures to avoid them.

5. How to Implement IoT Analytics in an Organization Efficiently

As IoT expands its reach to more industries, the demand for IoT data analytics roughly increases accordingly. A lot of companies are on their way to IoT adoption, but not all of them know how to implement it properly. What is the best way to implement IoT analytics within an organization so that it can be done efficiently? To help you better understand the procedure of IoT analytics implementation within an organization, we will lead you through some best practices that can result in a smooth and effective process.

- Determine the use cases: the first and foremost thing to do is to identify the specific use cases in which your organization can benefit from IoT data analytics. Once you have a clear understanding of your needs, you will be able to decide the appropriate approach and select the right IoT data analytics platform for your organization.
- Data collection: the next step is to set up a system for collecting raw data from various sources. This can be done by setting up and installing IoT sensors and other devices that can collect data about the different aspects of your business operations. In this stage, companies are often advised to leverage automation for data cleaning[1], as it can help to remove any invalid or incomplete data points and make the data more accurate and reliable.
- Data storage: after data has been collected, it is significant to store data in central data centers so that it can be accessed and analyzed when needed effortlessly. This can be done by using a cloud-based data storage platform.
- Data visualization: whether it is structured, unstructured, or semi-structured data, it needs to be visualized to make it easier and more comprehensive to understand and interpret later. At this point, you can make use of various data visualization tools to gain insights into your data.
- Data analysis: this is the core step of the entire process, in which data is analyzed to extract valuable insights. This can be done by using different types of data analytics tools and techniques, including various data analysis methods such as machine learning, predictive analytics, and statistical analysis.

IoT data analytics has been traveling so far since its inception, and it has become an integral part of many businesses. If you would like to make the most of your data assets and empower your business decisions, then it is time to embrace IoT data analytics. IoT data analytics can help businesses in a number of ways as long as you know how to do it right.

New Words

collect	[kə'lekt]	vt. 收集
valuable	['væljuəbl]	adj. 有价值的
actionable	['ækʃənəbl]	adj. 可操作的，可行动的
report	[rɪ'pɔːt]	n. 报告

---

1 Data cleansing or data cleaning is the process of detecting and correcting (or removing) corrupt or inaccurate records from a record set, table, or database and refers to identifying incomplete, incorrect, inaccurate or irrelevant parts of the data and then replacing, modifying, or deleting the dirty or coarse data.

occur	[ə'kɜː]	vi.	发生；出现
anomaly	[ə'nɒməli]	n.	异常，反常
uncover	[ʌn'kʌvə]	v.	发现，揭示
relationship	[rɪ'leɪʃnʃɪp]	n.	关系，联系
trend	[trend]	n. & vi.	倾向；趋势
model	['mɒdl]	n.	模型，模式
recommendation	[ˌrekəmen'deɪʃn]	n.	推荐；建议
subset	['sʌbset]	n.	子集
competitive	[kəm'petətɪv]	adj.	竞争的，有竞争力的
operational	[ˌɒpə'reɪʃənl]	adj.	运行的；操作的
inefficiency	[ˌɪnɪ'fɪʃntənsi]	n.	无效率，无能
malfunction	[ˌmæl'fʌŋkʃn]	vi.	失灵；发生故障
		n.	故障；功能障碍；失灵
wastage	['weɪstɪdʒ]	n.	消耗；浪费（量）；废物
preference	['prefrəns]	n.	偏爱
interest	['ɪntrəst]	n.	兴趣，爱好
hazard	['hæzəd]	vt.	冒险；使遭受危险
		n.	危险；冒险的事
construction	[kən'strʌkʃn]	n.	建筑物；建造
accident	['æksɪdənt]	n.	意外事件，事故
adoption	[ə'dɒpʃn]	n.	采用
appropriate	[ə'prəʊpriət]	adj.	适当的，恰当的，合适的
invalid	[ɪn'vælɪd]	adj.	无效的
incomplete	[ˌɪnkəm'pliːt]	adj.	不完整的，不完全的；未完成的
effortlessly	['efətləsli]	adv.	轻松地，不费力地
cloud-based	[klaʊdbeɪst]	adj.	基于云的
empower	[ɪm'paʊə]	vt.	使能够；授权；准许

Phrases

data analytic	数据分析
unstructured data	非结构化数据
business decision-making	业务决策，商务决策
data set	数据集
be broken down into…	被分解成……
descriptive analytics	描述性分析
historical data	历史数据
diagnostic analytics	诊断分析
drill down into	深入

data mining	数据挖掘
statistical analysis	统计分析
predictive analytics	预测分析
demand forecasting	需求预测，需求预报
prescriptive analytics	规定性分析
business outcome	业务成果
optimization algorithm	优化算法
big data	大数据
data source	数据源
real-time streaming data	实时流数据
edge computing	边缘计算
be classified into…	分（类）为……
asset utilization	资产利用
production line	生产线，流水线
personalized recommendation	个性化推荐
scale up	按比例增加，按比例提高
raw data	原始数据
data cleaning	数据清理
data center	数据中心
semi-structured data	半结构化数据
data visualization tool	数据可视化工具
data asset	数据资产

Analysis of Difficult Sentences

[1] As its name suggests, IoT data analytics is the act of analyzing data generated and collected from IoT devices by utilizing a specific set of data analytics tools and techniques.

本句中，generated and collected from IoT devices 是过去分词短语，作定语，修饰和限定 data。by utilizing a specific set of data analytics tools and techniques 是介词短语，作方式状语。As its name suggests 的意思是"顾名思义"。

[2] The historical data collected from devices are processed and analyzed to generate a report that describes what took place, when it occurred, and how often it did.

本句中，collected from devices 是一个过去分词短语，作定语，修饰和限定主语 The historical data。to generate a report that describes what took place, when it occurred, and how often it did 是一个动词不定式短语，作目的状语。在该不定式短语中，that describes what took place, when it occurred, and how often it did 是一个定语从句，修饰和限定 a report。

[3] Diagnostic analytics make use of techniques like data mining and statistical analysis to uncover hidden patterns and relationships in data that can offer actionable insights into the causes of specific problems.

本句中，to uncover hidden patterns and relationships in data that can offer actionable insights into the causes of specific problems 是一个动词不定式短语，作目的状语。在该不定式短语中，that can offer actionable insights into the causes of specific problems 是一个定语从句，修饰和限定 hidden patterns and relationships in data。make use of 的意思是"利用，使用"。

[4] And as a result, IoT analytics can be used to solve various issues and problems that cannot be addressed by big data analytics alone, such as real-time streaming data analysis, near-time processing, edge computing, predictive maintenance, etc.

本句中，to solve various issues and problems that cannot be addressed by big data analytics alone, such as real-time streaming data analysis, near-time processing, edge computing, predictive maintenance, etc.是一个动词不定式短语，作目的状语。在该短语中，that cannot be addressed by big data analytics alone 是一个定语从句，修饰和限定 various issues and problems。such as real-time streaming data analysis, near-time processing, edge computing, predictive maintenance, etc. 是对 various issues and problems 的举例说明。as a result 的意思是"因此"。

[5] This is because IoT data analytics can be deployed on the cloud, which allows businesses to pay only for the resources they use.

本句中，because IoT data analytics can be deployed on the cloud 是一个由 because 引导的表语从句，作 is 的表语，说明 This 的原因。This 指的是 With IoT data analytics, businesses can scale up their operations quickly and easily without incurring any additional costs。which allows businesses to pay only for the resources they use 是一个非限定性定语从句，对 IoT data analytics can be deployed on the cloud 进行补充说明。在该从句中，they use 是一个定语从句，修饰和限定 the resources。pay for 的意思是"支付"。

Exercises

【EX.1】**Answer the following questions according to the text.**
1. What Is IoT data analytics?
2. What does IoT analytics enable?
3. What is descriptive IoT analysis useful for?
4. What is predictive IoT analytics used to do?
5. What does diagnostic analytics do?
6. What is prescriptive IoT analytics?
7. What is the only distinction between IoT and big data analytics?
8. What are the business benefits of IoT analytics?
9. What are the technical benefits of IoT analytics?
10. What is it significant to do after data has been collected?

【EX.2】**Translate the following terms or phrases from English into Chinese and vice versa.**

1.	actionable	1.	
2.	adoption	2.	
3.	collect	3.	

4.	incomplete	4.	
5.	invalid	5.	
6.	model	6.	
7.	recommendation	7.	
8.	be classified into	8.	
9.	big data	9.	
10.	data analytic	10.	
11.	数据挖掘	11.	
12.	数据清理	12.	
13.	数据可视化工具	13.	
14.	边缘计算	14.	
15.	半结构化数据	15.	

【EX.3】Translate the following sentences into Chinese.

1. There are numerous benefits of IoT data analytics, for example, optimizing operational efficiency.
2. Data mining involves collecting information from data stored in a database.
3. Companies employ predictive analytics to find patterns in this data to identify risks and opportunities.
4. Diagnostic analytics looks at what has happened to try to determine the root cause of those events using mathematical functions such as probabilities, likelihoods, and the distribution of outcomes.
5. He can tell the difference between diagnostic analytics and prescriptive analytics.
6. If you enter any invalid information, there will be an error code displayed.
7. This particular model is one of our biggest sellers.
8. The data collected from IoT devices can provide valuable insights for the company.
9. There must have been a computer malfunction.
10. IoT data analytics can help businesses to reduce energy consumption.

【EX.4】Complete the following passage with appropriate words in the box.

incorporate	surf	networking	super-fast	mobile
connectivity	send	low-power	conform	cabling

　　Bluetooth and WiFi are both wireless networking standards that provide connectivity via radio waves. The main difference: bluetooth's primary use is to replace cables, while WiFi is largely used to provide wireless, high-speed access to the Internet or a local area network.

1. Bluetooth

First developed in 1994, bluetooth is a ___1___, short-range (30 feet) networking specification with moderately fast transmission speeds of 800 kilobits per second. Bluetooth provides a wireless, point-to-point, "personal area network" for PDAs, notebooks, printers, ___2___ phones, audio components, and other devices. The wireless technology can be used anywhere you have two or more devices that are bluetooth enabled. For example, you could ___3___ files from a notebook to a printer without having to physically connect the two devices with a cable.

A few notebooks, such as the IBM ThinkPad T30, now include built-in bluetooth ___4___. And $129 will buy you a bluetooth card for expansion-slot Palm PDAs, allowing you to connect to printers, notebooks, mobile phones, and other devices without cables.

Despite the promises of bluetooth, however, hardware makers have been slow to ___5___ it into their products. Some experts believe it could be eight years before Bluetooth is commonly used. They attribute the technology's lagging adoption rate to poor usability and confusion about what Bluetooth is and does.

2. WiFi

Short for wireless fidelity, WiFi is a user-friendly name for devices that have been certified by the Wireless Ethernet Compatibility Alliance to ___6___ to the industry-standard wireless networking specification IEEE 802.11b. WiFi began appearing in products in late 1998. The standard currently provides access to Ethernet networks such as a corporate LAN or the Internet at ___7___ speeds of up to 11 megabits per second.

WiFi connections can be made up to about 300 feet away from a "hot spot" (slang for a WiFi networking node). When your notebook or PDA has a WiFi networking card or built-in chip, you can ___8___ the Internet at broadband speeds wirelessly. WiFi networking nodes are proliferating globally; many Starbucks locations, for instance, offer access to WiFi hot spots for a fee.

Many notebooks today have IEEE 802.11b built-in; those that don't can be adapted via WiFi connectivity PC Cards. WiFi is also the basis for some home networking products, allowing you to share high-speed Internet connections without ___9___. Late last year, products featuring a newer wireless networking specification, IEEE 802.11a (called WiFi5 by WECA), debuted. This standard provides transmission speeds of up to 54 mbps. Wireless ___10___ is expected to grow in popularity as a practical, flexible way to replace some LANs. With wireless networking, for instance, workers can carry their notebooks from cubicle to conference room and stay connected to the corporate network.

【EX.5】 Translate the following passage into Chinese.

ZigBee

Pioneered by Philips Semiconductors, ZigBee is a low data rate, two-way standard for home automation and data networks. The standard originates from the Firefly Working Group and provides a specification for up to 254 nodes including one master managed from a single remote control. Real usage examples of ZigBee includes home automation tasks such as turning lights on, turning up the heat, setting the home security system, or starting the VCR. With ZigBee all these tasks can be done from anywhere in the home at the touch of a button. ZigBee also allows for dial-in access via the Internet for automation control.

The ZigBee standard uses small and very low-power devices to connect together to form a wireless control web. A ZigBee network is capable of supporting up to 254 client nodes plus one full functional device (master). ZigBee protocol is optimized for very long battery life measured in months to years from inexpensive, off-the-shelf non-rechargeable batteries, and can control lighting, air conditioning and heating, smoke and fire alarms, and other security devices. The standard supports 2.4 GHz (worldwide), 868 MHz (Europe) and 915 MHz (Americas) unlicensed radio bands with range up to 75 meters.

Text B

IIoT

With 5G, machine learning, artificial intelligence[1], and other new technologies available on the market to use today, we're now beginning the fourth industrial revolution (commonly named Industry 4.0).Industry 4.0 is a new era of the industrial revolution that focuses on connecting machines to databases, automating processes, machine learning, and real-time information. As Industry 4.0 develops, IIoT(Industrial Internet of Things) will play a significant role within the business setting and significantly impact how businesses operate. We'll see many organizations transform traditional manufacturing operations using IIoT to enhance production lines and increase profits.

1. What Is IIoT?

IIoT refers to the use of the IoT in industrial environments. It is a collection of linked sensors, machines and devices connected to computers' industrial applications. Organizations take advantage of superior data collection by connecting all these devices, sensors, computers, and machines. The collected data is then sent to a centralized cloud system, where it is exchanged with other devices and end users to improve productivity and efficiency levels.

The industrial internet of things (IIoT) uses machine learning, big data, real-time analytics, and machine-to-machine (M2M) communications to help corporations and enterprises create better processes for industrial development. Machines using IIoT technologies are better at capturing/analyzing data in real-time and communicating important information than traditional

---

1　Artificial intelligence or AI refers to the simulation of human intelligence in machines that are programmed to think and act like humans.

human processes that are often slow with biases and errors.

2. IIoT vs IoT

Different but somewhat similar, IIoT extends from the IoT technologies. IIoT brings IoT technologies into the industrial sector – hence the added "I" at the beginning.

IIoT leverages IoT technologies in the business and industrial sectors. Like optimizing your home with smart lights, connected thermostats, and other smart home devices, the industrial sector also uses IoT technologies to make their processes, insights, and outputs "smart".

Although both technologies use the same standard protocols, IIoT's goals, applications, scalability, and technical details differ from IoT's.

Compared to IoT's focus on consumer convenience and daily household tasks, IIoT's main priority is monitoring outputs. Industries using IIoT technologies want to reach maximum efficiency, productivity, and optimization.

Scalability is an essential difference between IoT and IIoT. IIoT is much more scalable than IoT, allowing thousands of new sensors to connect to various robots, machines, controllers, and other devices. Full integration makes IIoT successful in data collection and analyzing exchanges between devices, machines, and humans. IoT, however, only connects to a limited number of devices and sometimes may only connect within a specific line of products.

Because of IIoT's need to scale with the industry, it must be more technical and detailed than other IoT deployments. IIoT uses complex devices and systems to assist existing manufacturing, supply chain, and business models, meaning setup may need to be customized specifically to a business' current operations. Although customizable to an extent, IoT isn't as detailed and doesn't require a complex setup to start. Most IoT devices are easy to install and operate over your home's WiFi.

3. Main Parts of a Standard IIoT System

A standard IIoT system shares data between hundreds of thousands of devices and networks. Within the system, there are four main parts:
- External devices that can communicate and store information about themselves, their productivity, and workload (these could include sensors, robots, remote devices, computers, driverless vehicles, controllers, and more).
- A data communications infrastructure that allows data to be transferred between devices and the cloud (can be public and/or private).
- A central cloud system that stores all the data generated by the various IIoT devices.
- An application programming interface (API)[1] or other applications to create business information from raw data collected from IIoT devices.

In a nutshell, external devices are the devices that do the work and record data on themselves.

---

1 An API, or application programming interface, is a set of defined rules that enable different applications to communicate with each other.

The communications infrastructure then collects the data from the devices, and bring them to other external devices and the cloud. The cloud stores all the raw data. A worker then uses an application to pull the relevant data needed.

Although laid out simply here, attaining this network can be a challenge. With so much data being transferred between devices, the cloud, and people, it can overload a network without the right equipment and communications layout. It's also important to note that not every IIoT network looks the same. Businesses and organizations need customized networks based on their current goals, outputs, and product strategy.

4. Benefits of IIoT

4.1 Increased Productivity and Efficiency

One of the primary benefits of IIoT is increasing productivity. IIoT increases productivity through optimizing and automating processes already set in place. Using a variety of intelligent sensors, devices, controllers, and more, organizations can build systems to automate and speed up current processes. Increased productivity is manufacturing's focus – getting more products out onto the market faster.

4.2 Cost Savings

Although investing in an IIoT system can be an expensive up-front cost, it pays off in the long run. Organizations that invest in IIoT systems can enhance productivity/efficiency, increase quality control[1], manage inventories easier, reduce labour costs and save on energy bills making IIoT a robust investment.

4.3 Improved Agility

Productivity and efficiency are essential in an organization, but agility is crucial, especially as changes emerge. Incorporating IIoT into an organization along with big data, AI, and cyber manufacturing systems creates a flexible and scalable approach to improve agility.

IIoT's ability to collect, organize, and even predict data quickly makes it agile. Data from IIoT devices are collected in real-time and organized so decision-makers can decide on an outcome right away. Using IIoT within a system also provides decision-makers with predictive analytics allowing them to see what may happen in the future based on past trends.

4.4 Real-time Monitoring

Real-time monitoring in an industrial setting allows management to see what is happening right now. Compared to traditional monitoring (where data on a single day is sometimes not seen for a few days or even weeks), real-time monitoring allows management to make quick changes. There's no point in making decisions based on old data.

Monitoring efficacy in real-time with IIoT also allows managers to get a 360-degree view of how operations are running. Suppose a manager notices that a robot or controller isn't operating correctly. In that case, they can not only send someone to fix the problem but also understand

---

1 Quality control (QC) is a procedure or set of procedures intended to ensure that a manufactured product or performed service adheres to a defined set of quality criteria or meets the requirements of the client or customer.

how that affects the operation chain as a whole.

Other benefits also include:

- Predictive maintenance: big data and AI can predict when devices need to be maintained.
- Quality management: having an IIoT setup improves product quality management.
- Supply chain management: with real-time and big data, managing supply chains is more straightforward.
- Zero defects manufacturing: having robots and other devices on the production line reduces defects in manufacturing products.

5. IIoT Challenges

5.1 IIoT Data Integration Challenges

With massive sets of big and raw data being collected from thousands of sensors, combining all this data to create meaningful insights is a significant challenge for organizations. How can organizations collect and organize this data, and make decisions based on the data?

Big data analytics projects typically use extract, transform, and load (ETL)[1] – a traditional IT methodology. When asked to provide data intelligence from multiple systems, data systems architects extract the data and place it into a familiar location, then transform the data by normalizing it and cleaning it. Finally, they load it into a common site for decision-makers, hence ETL.

Unfortunately, ETL often stumbles within the significant complexities of IIoT. Since the data collected from IIoT is massive, traditional systems like ETL cannot handle it well, and ETL's manual processes and rigid architecture can't keep up with IIoT's speedy and customizable needs.

The reality is that IIoT requires automation and, in most cases, the application of AI to be fully optimized. AI will enable real-time alerts to the collected data and flag irregularities and potential issues as they arise.

5.2 IIoT Cyber Security Challenges

The sheer volume of critical and sensitive data[2] that IIoT produces makes it a large target for incoming cyber attacks. If a cyber attack were successful on an IIoT setup, the damage could be catastrophic.

Since IIoT may have thousands of different sensors, machines and other equipment installed on the system, there are plenty of opportunities to attack. It's easy for an attacker to take advantage of a simple mistake. Attackers aren't just from external, they may be from internal, making securing an IIoT system even more of a challenge.

---

1 Extract, transform, and load (ETL) is the process of combining data from multiple sources into a large, central repository called a data warehouse. ETL uses a set of business rules to clean and organize raw data and prepare it for storage, data analytics, and machine learning (ML).

2 Sensitive data is information stored, processed, or managed by an individual or organization that is confidential and only accessible to authorized users with proper permission, privileges, or clearance to view it.

The best cyber security plan is a predictive one. While implementing IIoT, do so with security top of mind.

New Words

transform	[træns'fɔːm]	vt.	改变，变换
corporation	[ˌkɔːpə'reɪʃn]	n.	公司；法人
bias	['baɪəs]	n.	偏差
insight	['ɪnsaɪt]	n.	洞察力，洞悉
productivity	[ˌprɒdʌk'tɪvəti]	n.	生产率，生产力
controller	[kən'trəʊlə]	n.	控制器
workload	['wɜːkləʊd]	n.	工作量，工作负担
overload	[ˌəʊvə'ləʊd]	vt.	使超载，超过负荷
	['əʊvələʊd]	n.	过量，超负荷
layout	['leɪaʊt]	n.	布局，安排；布置图，规划图
agility	[ə'dʒɪləti]	n.	敏捷性
decision-maker	[dɪ'sɪʒn'meɪkə]	n.	决策人
chain	[tʃeɪn]	n.	链条
defect	['diːfekt]	n.	瑕疵，欠缺
methodology	[ˌmeθə'dɒlədʒi]	n.	方法学，方法论
multiple	['mʌltɪpl]	adj.	多重的，多个的，多功能的
normalize	['nɔːməlaɪz]	vt.	使规范化，使标准化
stumble	['stʌmbl]	vi.	跌倒，蹒跚；弄错
customizable	['kʌstəmaɪzəbəl]	adj.	可定制的
flag	[flæg]	vt.	标示，标记
irregularity	[ɪˌregjə'lærəti]	n.	违规
catastrophic	[ˌkætə'strɒfɪk]	adj.	灾难的，惨重的
attacker	[ə'tækə]	n.	攻击者
plan	[plæn]	n.	计划，打算

Phrases

artificial intelligence	人工智能
fourth industrial revolution	第四次工业革命
automating process	自动处理，自动化过程
centralized cloud system	集中式云系统
real-time analytic	实时分析
human process	人工处理
industrial sector	工业部门，工业领域
line of product	产品系列
business model	企业模型，商业模式

driverless vehicle	无人驾驶车辆
lay out	展示；安排；陈设
speed up	（使）加速
up-front cost	前期成本
pay off	取得成功，带来好结果
quality control	质量管理
manage inventory	管理库存
labour cost	人工成本，人力成本
cyber manufacturing system	网络制造系统
zero defects manufacturing	零缺陷制造
data systems architect	数据系统架构师
manual process	手动处理
rigid architecture	刚性结构
keep up	保持
sensitive data	敏感数据
plenty of	很多，大量

Abbreviations

IIoT (Industrial Internet of Things)	工业物联网
5G (Fifth Generation)	第五代
API (Application Programming Interface)	应用程序编程接口
ETL (Extract, Transform, and Load)	提取、转换和加载

Exercises

【EX.6】 Answer the following questions according to the text.

1. What is Industry 4.0?
2. As Industry 4.0 develops, what will IIoT do?
3. What does IIoT refer to? What is it?
4. What are machines using IIoT technologies better at?
5. What do industries using IIoT technologies want to do?
6. How many main parts are there within a standard IIoT system? What are they?
7. What is one of the primary benefits of IIoT? How does IIoT do it?
8. What does monitoring efficacy in real-time with IIoT also allow managers to do?
9. What do other benefits of IIoT also include?
10. What makes it a large target for incoming cyber attacks?

【EX.7】 Translate the following terms or phrases from English into Chinese and vice versa.

1. bias   1. _____
2. agility   2. _____

3.	controller	3.	
4.	flag	4.	
5.	overload	5.	
6.	transform	6.	
7.	artificial intelligence	7.	
8.	automating process	8.	
9.	quality control	9.	
10.	real-time analytic	10.	
11.	敏感数据	11.	
12.	*n.* 计划，打算	12.	
13.	*n.* 生产率，生产力	13.	
14.	*adj.* 多重的，多个的，多功能的	14.	
15.	*vt.* 使规范化，使标准化	15.	

【EX.8】Translate the following sentences into Chinese.

1. Good teachers have insight into the problems of students.
2. Increased productivity is accompanied by a progressive decrease in production costs.
3. The controller's job is to control the application flow.
4. In each control interval, only one controller is executed.
5. Signal processing technique can effectively reduce system workload.
6. Using XML for the agency of data exchange has brought infinite agility.
7. The system is extensible and can provide individualized customizable software service.
8. I've flagged the paragraphs that we need to look at in more detail.
9. Careful contrast of the two plans shows some important differences.
10. The Windows allow a computer user to execute multiple programs simultaneously.

Reading Materials

Artificial Intelligence of Things (AIoT)

Artificial Intelligence of Things[1] (AIoT) emerges as a game-changing force in the rapidly evolving world of technology, enabling businesses to optimize operations, enhance customer experiences, and drive unprecedented[2] innovation.

1. What Is AIoT?

AIoT refers to the powerful fusion of Artificial Intelligence (AI) and the Internet of Things (IoT).

1　Artificial Intelligence of Things：人工智能物联网
2　unprecedented [ʌnˈpresɪdentɪd] *adj.* 前所未有的，空前的

In simple terms, AI concepts, algorithms, and technologies are blended with[1] smart devices, sensors, and everyday objects that are connected to the internet. This merging technology enables these connected devices to collect and analyze vast amounts of data, make intelligent decisions and take actions autonomously[2] without human intervention.

2. How Does AI Enhance IoT Capabilities?

AI infuses intelligence into[3] a vast network of interconnected devices to enhance the capabilities of IoT, changing the way these devices are used to collect, process, and utilize data.

Here's a concise breakdown of how AI empowers IoT.

2.1 Advanced Data Analytics

AI empowers[4] IoT devices with robust data analytics capabilities. This allows them to seamlessly process and analyze vast volumes of data in real-time, leading to enhanced precision in insights. Thus, it becomes easier for them to produce high-quality data quickly and take prompt action.

2.2 Predictive Analytics

AI algorithms have the capability to predict future events and trends by analyzing historical data. This becomes particularly relevant in the context of IoT as it empowers devices to anticipate potential issues. This way, users can understand the risks and take proactive measures in no time before anything bad happens.

2.3 Machine Learning (ML)

By utilizing machine learning, the IoT devices can continuously acquire knowledge from data and enhance their performance over time. This inherent[5] self-optimization process results in heightened efficiency and accuracy.

2.4 Natural Language Processing (NLP)[6]

AI empowers IoT devices to comprehend and address human commands and inquiries through natural language processing (NLP). This cultivates smooth interactions and facilitates user-friendly experiences.

2.5 Resource Management

AI's predictive capabilities can help you effectively manage resources in your IoT devices. This helps ensure optimized resource consumption and minimal waste.

2.6 Security and Privacy Tools

AI enhances security in the IoT realm by integrating tools and technologies that can identify potential vulnerabilities and actively safeguard against cyber threats. It includes modern security

---

1 be blended with: 与……混合
2 autonomously [ɔː'tɒnəməsli] *adv.* 自主地
3 infuse ... into ...: 把……注入……
4 empower [ɪm'paʊə] *vt.* 授权；准许
5 inherent [ɪn'hɪərənt] *adj.* 固有的，内在的
6 Natural Language Processing (NLP): 自然语言处理

and privacy technologies like 2-factor authentication, single sign-on[1], etc. This proactive[2] approach ensures a robust defense system.

3. Technologies Used in AIoT

3.1 Machine Learning

Machine learning (ML) can analyze a vast data set to enable effective decision-making. Also, ML algorithms enable IoT devices to recognize patterns, trends, and anomalies[3]. In addition, ML enables predictive maintenance, real-time analytics, and personalized user experiences.

3.2 Edge Computing

Edge computing is a highly crucial concept in the AIoT landscape. It enables the processing of data directly on IoT devices or edge servers, closer to where the data originates, instead of sending everything to central cloud servers. This approach significantly reduces latency[4], enhances real-time response, and minimizes bandwidth usage.

3.3 AIoT Platforms and Frameworks

AIoT platforms and frameworks are helpful tools in building and deploying AIoT applications. They offer prebuilt[5] AI models, data processing pipelines, and connectivity solutions that simplify the development process.

Examples of popular AIoT platforms include Google Cloud IoT, Microsoft Azure IoT, and AWS IoT. These platforms provide a wide range of services to help integrate AI with IoT devices quickly.

4. AIoT Applications for End Users

4.1 Smart Home

Through the power of AIoT, homes are transformed into havens where devices seamlessly communicate, adapt, and cater to your preferences.

Take smart thermostats like Nest, for example. They learn your temperature preferences and automatically adjust settings for both comfort and energy efficiency.

Additionally, voice-controlled assistants such as Amazon Echo or Google Home provide you with the convenience of controlling lights, appliances, and even playing your favorite music through simple voice commands.

4.2 Industrial Internet of Things (IIoT)

Within the manufacturing sector, IoT sensors gather real-time data from machines, while AI algorithms anticipate[6] maintenance requirements. Consequently, this reduces downtime and optimizes productivity levels.

---

1 single sign-on：单点登录

2 proactive [ˌprəʊˈæktiv] *adj.* 积极主动的，先发制人的

3 anomaly [əˈnɒməli] *n.* 异常，反常；不规则

4 latency [ˈleɪtənsi] *n.* 延迟

5 prebuilt [priːˈbɪlt] *adj.* 预置的，预建的

6 anticipate [ænˈtɪsɪpeɪt] *vt.* 预测，预感，预见

4.3 Smart Cities

AIoT is revolutionizing the way cities function, making them smarter and more sustainable. For example, intelligent traffic management systems harness[1] the power of AI through cameras and sensors to optimize the flow of traffic and effectively reduce congestion and travel time.

4.4 Healthcare and Wearables

Thanks to A IoT, the healthcare industry has received a significant boost. For example, wearable devices are used to monitor vital signs[2], sleep patterns, and physical activities. These incredible[3] wearable devices can seamlessly transmit data to smartphones or cloud platforms.

4.5 Agriculture

AIoT revolutionizes precision agriculture to empower farmers to optimize their practices for increased yields and reduced resource wastage.

By deploying IoT sensors that monitor crucial factors like soil moisture, temperature, and humidity, alongside AI algorithms that meticulously[4] analyze the collected data, precise irrigation schedules can be determined.

Additionally, drones[5] equipped with advanced AI and IoT capabilities can efficiently survey large fields, promptly identify crop diseases, and facilitate targeted treatments.

5. AIoT Applications in Organizations

5.1 Predictive Maintenance

To effectively monitor machinery and equipment in real-time, you can implement AIoT. AI algorithms can utilize sensor data to anticipate maintenance requirements, leading to decreased downtime and preventing expensive breakdowns.

5.2 Enhanced Customer Experiences

AIoT empowers organizations to collect and analyze customer data from connected devices. This enables personalized products, services, and tailored[6] marketing efforts for a seamless and delightful customer experiences.

5.3 Energy Management

If your organization consumes a significant amount of energy, AIoT can help optimize the usage by intelligently managing the lighting, heating, and cooling systems based on factors like occupancy and weather conditions. This smart control results in notable[7] energy savings.

---

1 harness ['hɑːnɪs] *vt.* 利用
2 vital sign：生命体征
3 incredible [ɪnˈkredəbl] *adj.* 不可思议的；难以置信的
4 meticulously [məˈtɪkjələsli] *adv.* 异常细致地，无微不至地
5 drone [drəʊn] *n.* 无人机
6 tailored [ˈteɪləd] *adj.* 订做的
7 notable [ˈnəʊtəbl] *adj.* 值得注意的；显著的

5.4 Security and Surveillance[1]

AIoT can enhance organizational security through the utilization of intelligent video analytics, facial recognition technology, and anomaly detection techniques[2]. This powerful combination ensures the protection of valuable assets and personnel.

5.5 Smart Workplaces

Did you know that AIoT has the power to transform workplaces into smart environments? By seamlessly integrating IoT devices and AI-powered systems, organizations can experience automated office management, improved employee productivity and optimized resource utilization.

6. Advantages of AIoT

6.1 Enhanced Efficiency

AIoT enhances processes by analyzing real-time data from interconnected devices. This data-driven approach enables you to predict maintenance needs, minimize downtime, and maximize efficiency. Furthermore, through task and decision-making automation, AIoT significantly boosts productivity, saving time and valuable resources.

6.2 Smart Decision-making

AIoT systems can handle immense[3] data volumes from various sources and deliver valuable insights that empower you to make better, informed decisions[4]. By utilizing advanced analytics, businesses can develop a deeper understanding of customer behavior, market trends, and operational patterns. This knowledge leads to improved strategies and outcomes for your business.

6.3 Personalization and Customer Experience

AIoT offers personalized experiences by comprehending users' preferences and behaviors. Through utilizing smart devices, AIoT can customize services according to specific needs and provide personal recommendations, therefore, enhancing customer-engagement and satisfaction[5]. satisfaction[5].

6.4 Improved Safety and Security

AIoT enhances safety by monitoring environments and predicting potential hazards[6]. It provides real-time anomaly detection that enables quick responses to emergencies and prevents accidents. Moreover, AI-driven cyber security ensures the protection of interconnected devices from cyber threats, guaranteeing data privacy and integrity.

1　surveillance [sɜːˈveɪləns] *n.* 监督，管制
2　anomaly detection techniques：异常检测技术
3　immense [ɪˈmens] *adj.* 极大的，巨大的
4　informed decision：知情决策
5　satisfaction [ˌsætɪsˈfækʃn] *n.* 满意度
6　hazard [ˈhæzəd] *n.* 危险

6.5 Reduced Environmental Impact

AIoT is vital for promoting sustainable[1] practices. With its ability to optimize energy consumption, manage waste effectively, and allocate resources efficiently, AIoT plays a crucial role in reducing the environmental impact of industries and smart cities.

7. Challenges of AIoT

7.1 Data Security and Privacy

The influx of data from interconnected devices can bring significant security and privacy risks. To safeguard user data, AIoT systems must handle sensitive information with care. This involves preventing unauthorized access, data breaches, and misuse through robust encryption, authentication, and access control mechanisms.

7.2 Interoperability and Standardization

In the vast realm of IoT, a wide array of devices and technologies coexist[2], but this diversity can sometimes give rise to compatibility issues. When communication protocols lack standardization, it may cause inefficiencies[3] in smooth integration and data exchange among AIoT components.

7.3 Complex Integration

Integrating AI capabilities into existing IoT infrastructures can pose challenges for organizations. Modifying legacy systems to effectively accommodate AI algorithms can be difficult. So, ensuring a smooth integration process is necessary to unlock the full potential of AIoT and derive valuable insights from collected data.

7.4 Reliability and Trust

AIoT decisions significantly impact critical systems, including healthcare and autonomous vehicles. It is crucial to prioritize[4] the reliability of AI algorithms and establish trust in the decision-making process.

7.5 Power Consumption and Sustainability[5]

AIoT devices often operate on limited power sources, such as batteries. But balancing the computational demands of AI algorithms with energy-efficient IoT devices can be quite challenging.

So, you must put effort into optimizing power consumption. You can explore sustainable energy solutions to extend the longevity of A IoT deployments and minimize their environmental impacts.

1　sustainable [sə'steɪnəbl] *adj.* 可持续的
2　coexist [ˌkəʊɪɡ'zɪst] *vi.* 同时共存
3　inefficiency [ˌɪnɪ'fɪʃənsi] *n.* 无效率
4　prioritize [praɪ'ɒrətaɪz] *vt.* 按重要性排列，划分优先顺序
5　sustainability [səˌsteɪnə'bɪlɪti] *n.* 持续性

参考译文

Text A 物联网数据分析

1. 什么是物联网数据分析？

顾名思义，物联网数据分析是利用一组特定的数据分析工具和技术来分析从物联网设备生成和收集的数据的行为。物联网数据分析背后的真正思想是将来自异构物联网生态系统中各种设备和传感器的大量非结构化数据转换为有价值且可操作的见解，以推动合理的业务决策和进一步的数据分析。此外，物联网分析可以识别数据集内的模式，包括当前状态和历史数据，可用于对未来事件进行预测和调整。

2. 不同类型的物联网数据分析

由于进行物联网分析的目的是收集服务于不同目的的见解，因此它可以分为四种主要类型。

2.1 描述性分析

描述性分析主要关注过去发生的事情。对从设备收集的历史数据进行处理和分析，生成一份报告，描述发生的事情、事情发生的时间及频率。这种类型的物联网分析对于提供有关物或人的行为的特定问题的答案非常有用，也可用于检测任何异常情况。

2.2 诊断分析

与描述性分析不同，诊断分析通过深入研究数据以确定特定问题的根本原因，进一步回答为什么会发生某些事情。诊断分析利用数据挖掘和统计分析等技术来揭示数据中隐藏的模式和关系，从而为特定问题的原因提供可操作的见解。

2.3 预测性分析

顾名思义，预测性分析用于通过分析历史数据和趋势来预测未来事件。这种类型的分析利用各种统计方法和机器学习算法来构建可用于预测未来事件的模型。此类分析在支持与库存管理、需求预测等相关的业务决策方面发挥着重要作用。

2.4 规范性分析

规范性分析是最先进的物联网分析类型，它不仅可以预测未来会发生什么，还可以建议应采取的措施来获得所需业务成果。这种类型的分析利用优化算法来确定实现特定目标应采取的最佳行动方案。

3. 物联网与大数据分析之间的关系

说到海量数据就会想到大数据分析，它们之间有什么联系？事实上，人们经常发现物联网分析和大数据分析相互混淆。它们之间唯一的区别是数据源：大数据分析处理来自广泛流和来源的数据集，而物联网分析仅收集和分析连接的物联网设备和传感器生成的数据。因此，可以说物联网分析是大数据分析的一个子集，有助于理解来自物联网生态系统中连接设备的数据。物联网分析可以用来解决仅靠大数据分析无法解决的各种问题，如实时流数据分析、近实时处理、边缘计算、预测性维护等。因此，物联网分析和大数据分析的结合可用于获得竞争优势并推动业务价值。

4. 物联网分析的优势

物联网分析的优势颇多，可以主要分为两大类：商业优势和技术优势。

4.1 物联网分析的商业优势

- 优化运营效率：通过分析物联网设备生成的数据，企业可以识别导致效率低下的问题，然后采取措施解决这些问题。例如，食品和饮料公司可以使用物联网数据分析来实时跟踪冰箱的温度，并防止由于停电或设备故障而导致食物变质。
- 降低成本：物联网分析可以通过多种方式帮助企业节省资金，如减少能源消耗、最大限度地减少停机时间及提高资产利用率。例如，制造公司可以使用物联网分析来监控其生产线的运行并进行调整以避免材料浪费。
- 增强客户体验：物联网分析可用于收集和分析客户数据，以了解他们的需求和偏好。反过来，这可以帮助企业设计出更好的产品和服务来满足客户的需求。例如，零售商可以使用物联网分析来跟踪顾客在其商店中的动向，然后根据他们的兴趣提供个性化推荐。
- 提高安全性：通过分析来自各种传感器的数据，企业可以识别潜在的安全隐患并采取预防措施。例如，建筑公司可以使用物联网分析来监控其设备和机械的状况，以避免发生事故。

4.2 物联网分析的技术优势

- 实时数据分析：物联网分析的主要优势之一是其分析实时数据点的能力。这是可能的，因为使用了流分析，此类分析可以在生成数据时处理数据。
- 提高可扩展性：通过物联网分析，企业可以快速、轻松地扩大运营规模，而不会产生任何额外成本。这是因为物联网分析可以部署在云上，这使得企业只需为他们使用的资源付费。
- 提高准确性：物联网分析可以帮助企业实现数据分析的高度准确性。这是因为物联网分析可用于从大量来源收集数据，然后使用先进的分析技术对其进行分析。
- 增强安全性：物联网分析还可以帮助企业提高数据的安全性。这是因为物联网分析可用于识别和跟踪潜在威胁，然后采取措施避免它们。

5. 如何在组织中高效实施物联网分析

随着物联网扩展到更多行业，对物联网分析的需求也相应增加。许多公司都在采用物联网，但并非所有公司都知道如何正确实施。在组织内实施物联网分析以使其高效完成的最佳方法是什么？为了帮助更好地了解组织内物联网分析实施的过程，将引导完成一些最佳实践，从而实现平稳有效的流程。

- 确定用例：首先要做的也是最重要的事情，即确定组织可以从物联网分析中受益的特定用例。一旦清楚地了解需求，将能够决定适当的方法并为组织选择正确的物联网数据分析平台。
- 数据收集：下一步是建立一个从各种来源收集原始数据的系统。这可以通过设置和安装物联网传感器及其他可以收集有关业务运营不同方面的数据的设备来完成。在这个阶段，通常建议公司自动化清理数据，因为可以帮助删除任何无效或不完整的数据点，并使数据更加准确和可靠。

- 数据存储：收集完数据后，将数据存储在中央数据中心非常重要，这样可以在需要时访问和分析数据。这可以通过使用基于云的数据存储平台来完成。
- 数据可视化：无论是结构化数据、非结构化数据还是半结构化数据，都需要可视化，这样就可以更容易、更全面地理解和解释数据。此时，可以利用各种数据可视化工具来深入了解数据。
- 数据分析：这是整个流程的核心步骤，通过分析数据来提取有价值的见解。这可以通过使用不同类型的数据分析工具和技术来完成，包括机器学习、预测分析和统计分析等数据分析方法。

自从有了物联网分析，它就一直被使用，并已成为许多企业不可或缺的一部分。如果想充分利用数据资产并为业务决策提供支持，那么是时候采用物联网分析了。只要知道如何正确实施，物联网分析就可以通过多种方式帮助企业。